PHYSICS OF THE STOICS

S. SAMBURSKY

Hebrew University, Jerusalem

PHYSICS

OF THE STOICS

Princeton University Press

Princeton, New Jersey

*Published by Princeton University Press, 41 William Street,
Princeton, New Jersey 08540*

Copyright © 1959 by S. Sambursky
All rights reserved

First Princeton Paperback printing, 1987
LCC
ISBN 0-691-08478-5
ISBN 0-691-02412-X (pbk.)
*Reprinted by arrangement with Routledge & Kegan Paul Ltd., Great
Britain*

*Printed in the United States of America by Princeton University
Press, Princeton, New Jersey*

PREFACE

THIS book gives an account of the physical doctrine of the
Stoics, all the aspects of which were based on a rigorous
continuum conception. The main outlines of Stoic physics were
briefly given in my earlier book *The Physical World of the
Greeks*, especially in chapter VI (The World of the Continuum)
and chapter VII (The Interdependence of Things).

In the present study I have tried to give a more detailed exposition of Stoic physics. The origins of some of its basic notions
are traced back to pre-Socratic sources, and the central concepts
such as pneuma, force, determinism, the infinite, together with
the beginnings of functional thought, have been more fully
analysed. In some respects, as for instance in the case of the Stoic
idea of Time, I had to revise my former contentions as a result of
further investigations. In others, such as their establishment of
the causal law and the distinction between partial and closed
systems, I have introduced new material.

In addition to quotations from original texts in the main body
of the book, further texts in translation illustrating each chapter
will be found in the Appendix.

My sincere thanks are due to Professor S. Pines of the Hebrew
University for his helpful criticism; to Mr. D. Leslie, M.Litt.,
who corrected the style; to Mr. J. Landau, M.A., who checked
my translations against the original Greek and to my wife who
prepared and typed the manuscript.

INTRODUCTION

HISTORIANS of the scientific thought of ancient Greece have paid considerable attention to the atomic School and have analysed the origins, basic notions and trends of the theories of Leucippos, Democritos and Epicuros and also investigated their influence on later generations. Among the reasons for the prominence which has been given to the Greek atomic theory, the three most important are the following: the striking similarity between some of the Greek concepts, based entirely on intuition, and those of the modern theory which have been confirmed by experience; the conclusions which the Greeks drew from their assumptions by applying methods of scientific inference; and finally, the appeal which the epistemology of the ancient atomists has for the scientists of our own atomic age.

The fact that atomism has gradually come into the foreground during the last hundred years must certainly be counted among the main reasons for the neglect of Stoic physics, a physics entirely based on the continuum concept. Even more important is the small number of extant Stoic writings and the low esteem in which some of their teachings were held by prominent historians of Greek philosophy, such as Prantl and Zeller.[1] The scope of Stoic contributions in the field of logic and the significance of their advance from the standpoint of Aristotle have become common knowledge with the work of Lukasiewicz and Mates. Although physics played a much less important role than logic in the Stoic philosophy, they developed a highly original and consistent system of physical concepts and applied it to the whole body of their teachings. In these we find anticipated basic ideas which have governed physical thought since the seventeenth century, and in spite of the fragmentary nature of the relevant sources, the outlines of a deeply conceived and well-elaborated continuum theory applied to the physical world are clearly evident. The essential feature of their theory is the dynamic notion of the concept of continuity which makes the Stoic doctrine one of the great original contributions in the history of physical systems, transcending by its implications the boundaries

[1] Cf. Mates, *Stoic Logic* (Berkeley, 1953), pp. 86 ff.

of pure physical thought and anticipating in many respects the approach to continuity which dominated the scientific ideas of Descartes, Huygens, Faraday and Maxwell. The dynamic continuum not only had profound influence on epistemology in the Hellenistic period, but it also moulded some of the basic ideas in the later teachings of Stoic ethics. Moreover, it led the older Stoa already to a first grasp of the modern mathematical notions of the function and the limit and so constitutes the first break through of the barriers of the merely static contemplation of mathematical quantities.

The aim of this investigation is to describe the main aspects of the Stoic continuum theory, to trace the origins of some of its notions back to pre-Stoic science and philosophy, and to show the attempts of the Stoics to work out a coherent, consistent system of thought which would explain the essential phenomena of the physical world by a few basic assumptions. It is of special interest to study the formation of a physical theory dating back to an age where systematic experimentation and mathematization of physical concepts were virtually unknown. Today, in the age of technology and mathematical physics, we are apt to underestimate the share of speculative and hypothetical elements in a scientific theory, and to overlook the logic inherent in certain conceptual pictures and analogies, which to a large extent establish a bridge that links the abstract elements of the theory with the brute facts provided by experience. The atomic theory of Leucippos, Democritos and Epicuros is one conspicuous example of this role of intuition in the process of theory formation; the continuum theory of the rival Stoic School is the second, a not less impressive and in many respects a more interesting one, especially in view of the fact that the notion of force and its field of action, omitted in the Epicurean world, was incorporated into the Stoic picture. The limitations of a one-sided continuous picture of the world are obvious, as are those of a one-sided atomistic one, and the lack of recourse to rigorous experimental tests in Greek science sets a fundamental limit to any comparison of either doctrine with modern theories as far as the description of observational details is concerned. But it is hardly possible to overrate the importance of the Stoic physical conceptions as the prototype of a mode of scientific approach to physical reality which up to now has proved indispensable in every attempt to give a coherent interpretation of Nature.

INTRODUCTION

Seen from the point of view of the history of ideas, what matters in Stoic physics and what remains of lasting interest is its continuum aspect. For this reason such details of the physical teachings of the Stoa which are of no relevance to this theory are omitted or dealt with only superficially in this study, especially those views which were taken over from other philosophies, particularly that of Aristotle. On the other hand, it is of course of importance to trace back the origins of some characteristic features of the continuum concept to earlier doctrines, such as the theory of total mixture to Anaxagoras, or the specific dynamic function of air and fire to Diogenes of Apollonia and Heracleitos.

The main body of the continuum theory was developed already by the older Stoics, particularly by Zeno and Chrysippos, i.e. in the fourth and third centuries B.C. The most important later addition to the doctrine was made in the first half of the first century B.C. by Poseidonios in his cosmic interpretation of the pneuma and the concept of sympathy. From the first century B.C. on, the centre of gravity of Stoic teachings shifted decidedly to ethics, and as far as is known today, no further contribution worth mentioning was made by the later Stoa in the field of physical thought. The scarcity of sources hardly allows any attempt to differentiate between the views of the older Stoa with regard to the various aspects of the continuum concept. Only a small number of clearly defined conceptual developments are recognizable in the physical teachings of the older Stoa which have been assigned by later authors to Zeno or Chrysippos or their pupils. In the writings of later generations, in Cicero, Plutarch, Plotin and the Aristotelian commentators, we very often find comments and criticisms directed against the Stoic physical doctrine as such without reference to a specific author.

The title of the present study refers therefore to the whole body of physical teachings which, according to our knowledge, was developed in a period roughly determined by the years 300 B.C. and 50 B.C. It can scarcely be doubted that the peak of that period coincides with Chrysippos and that he must be credited, if not with the lion's share of contributions towards Stoic physics, at least with the merit of having given its most detailed elaboration. It is worth while to remember in this connection that Chrysippos was a contemporary of two of the greatest scientists of Greek antiquity—Archimedes and Eratosthenes—of whom the latter is known to have been associated with the Stoa. The conjecture of a direct communication between Chrysippos and the prominent scientists of his time, although not proved, seems legitimate. Some passages in Chrysippos' fragments and the use of certain scientific termini possibly indicate his acquaintance with Archimedes' writings, at least.

It cannot be stressed too much how lamentably few quotations from Stoic writings have been preserved for us. We owe most in this respect to Plutarch, who in his essays on the Stoic teachings and in some other books of his *Moralia* here and there quotes a sentence or even a longer passage, mostly from some of the prolific writings of Chrysippos. When reading them one cannot get rid of the agonizing thought how enormously richer our insight into Stoic philosophy and indeed into the whole Hellenistic era would have been had the works of the Stoics survived to the same extent as those of Plato and Aristotle.

It goes without saying that Plutarch's choice of extracts was directed by a strong bias against the Stoa. Consequently the picture we are apt to get from them, taken as they are out of their context, is certainly somewhat one-sided. On the other hand, the descriptions given by many authors such as Cicero, Seneca, Diogenes Laertios, Simplicios of the physical theories of the great Stoic authors, detailed and exhaustive as they are, are partly subject to the same bias, and on the whole cannot be but poor substitutes for the original texts to anyone beset with the task of reconstructing the full meaning of a theory in which terminology and careful definitions played so significant a role.

CONTENTS

xi

Author's Note

In using Greek endings for Greek names I followed the example of the late G. Sarton without, however, any claim to consistency.

I

THE DYNAMIC CONTINUUM

1. *Pneuma and Coherence*

ACCORDING to the Stoic conception, the cosmic scene of material events, including conglomerate matter as well as space between bodies, is made up of a continuous whole. Like Aristotle, the Stoics exclude emphatically any possible existence of a void within the cosmos.[1] However, their cosmos is, in contradiction to that of Aristotle, an island embedded in an infinite void.[2] The cosmos is filled with an all-pervading substratum called *pneuma*, a term often used synonymously with *air*.[3] A basic function of the pneuma is the generation of the cohesion of matter and generally of the contact between all parts of the cosmos. The term coherence was originally used by Aristotle to express continuity in an essentially geometrical and topological sense,[4] but the Stoics gave it the physical and dynamic significance of cohesion within the physical world.

The property attributed to the pneuma of producing coherence can be traced back to pre-Socratic sources. This is of special interest for the study of the development of physical concepts in

---

[1] Diog. Laert., VII, 140; Cleom., *De motu circulari*, I, 1 (ed. Ziegler), p. 8.
[2] Plut., *De Stoic. repugn.*, 1054 b, says that this theory was repeatedly mentioned in Chrysippos' writings.
[3] e.g. Plut., loc. cit., 1053 f.
[4] Arist., *Metaph.*, 1069 a 6; cf. also *Metaph.*, 1061 a 33. See p. 5 in connection with *Metaph.*, 1015 b 36. For pre-Socratic origins of συνέχεια, cf. Parmenides (Diels, 28 B 8, 25): τῷ ξυνεχὲς πᾶν ἐστιν, ἐὸν γὰρ ἐόντι πελάζει.

ancient Greece, as it brings to light the biological origin of some of the basic physical notions. In a well-known fragment of Anaximenes it says: "As our soul, being air, holds us together, so do breath and air surround the whole universe."[5] Sextus Empiricus gives us the reason on which the Pythagoreans and Empedocles based their belief that the fellowship of men is not only with one another and with the gods, but also with animals: "For there is one spirit (*pneuma*) which pervades, like a soul, the whole Universe, and which also makes us one with them."[6]

Other evidently biological roots of the concept of pneuma can be found in Aristotle's exposition of the doctrine of the soul in the teachings of the Pythagoreans and other early philosophers.[7] Respiration is a characteristic sign of life, a fact which throws into relief the connection between soul (*pneuma*) and movement. The dynamic properties of the pneuma will be discussed later on; its identification with breath and soul in living organisms led to the Stoic hypothesis of the composition of the pneuma which was supposed to be a mixture of air and fire.[8] This composition was apparently made plausible by analogy of the pneuma with the warm puff of breath. As air represents the principle of Cold, the warmth of the human body makes it likely that the stuff souls are made of is a mixture of cold and hot, of air and fire. Galen[9] tells us that this mixture is so proportioned that it protects the living organism from extremes of temperature. Here we see clear associations with Alcmaeon's equilibrium theory of health[10] and with Hippocratic ways of thinking.[11] The passage in Galen just mentioned shows that the Stoics expanded the idea of mixture by adding the other two qualities of Dry and Moist, thus making possible the differentiation between the pneuma of the soul (*psyche*) and that of the world of plants (*physis*). The former is dry and warm whereas the pneuma of physis is moist and cold. We shall see later how important this principle of differentiation was, when it became the basis for the Stoic explanation of physical phenomena.

[5] Diels, 13 B 2.
[6] Sext. Emp., *Adv. math.*, IX, 127.
[7] See especially *De anima*, 404 a 21: "Movement is that which is closest to the nature of soul."
[8] e.g. Alex. Aphr., *De anima*, 26, 16; *De mixt.*, 224, 15.
[9] Galen, *De anim. mor.* (Arnim, II, 787); cf. also Arnim, II, 446.
[10] Diels, 24 B 4.   [11] *Ancient medicine*, XVI.

Another aspect of both components of the pneuma is their activity. Aristotle divided the four elementary qualities of matter into active—warm and cold, and passive—dry and moist and regarded the four elements as a result of the four physically possible combinations of these four qualities.[12] The Stoics who attribute one quality only to each of the four elements, defined Air and Fire as active, and Earth and Water as passive elements.[13] It might seem paradoxical that activity is attributed to the cold element (Air) as Cold was usually regarded as a rather passive quality.[14] The confusion which is found in this respect in Aristotle's writings[15] prevails also in later literature. One of the main reasons for this seems to be the constant confusion of air and water vapour. Whereas Poseidonios makes moisture responsible for the coldness of air over marshy ground,[16] his pupil Cicero stresses the caloric content of air: "Air must be deemed to be a sort of vaporized water, and this vaporization is caused by the motion of the heat contained in the water."[17] On the other hand, Plutarch points to the active participation of air during the freezing of water.[18] Making use of dialectic arguments, Plutarch finally assigns to air an intermediate position between fire and water.[19] Thus, "neutralization" of air emphasizes at the same time its active role in thermic processes.[20] Elasticity, linked with pressure, is another aspect of the active character of air; it was known in Greece already by the fifth century B.C. and is repeatedly mentioned by Aristotle.[21]

The function of fire as an active element is more obvious. Here, too, the starting-point was the observation of biological processes. "Innate heat", an expression coined by Hippocrates, was thought by Galen to be the cause of metabolism.[22] The often-quoted Stoic sentence, "Nature is an artistically working fire, going on

[12] Arist., *De gener. et corr.*, 329 b 25.
[13] Galen, *De nat. facult.*, I, iii, 8; Nemes., *De nat. hom.*, ch. 5.
[14] Arist., *Meteor.*, 382 b 5.
[15] I. Duering, *Aristotle's chemical treatise Meteorologica Book IV* (Goeteborg 1944). p. 81.
[16] Plut., *De primo frigido*, 951 f.
[17] Cicero, *De nat. deor.*, II, 27.
[18] Plut., loc. cit., 949 b:
[19] loc. cit., 951 e.
[20] The same process of neutralization can be seen with regard to the levity of air. Cf. below, p. 6.
[21] Arist., *Probl.*, XVI, 8; XXV, 1.
[22] Galen, *De nat. facult.*, I, xi, 25; II, iv. 89.

its way to create",[23] also points to the significance of heat in organic nature. On the other hand, one can infer from the passage in Cicero's *De natura deorum* dealing with the nature of heat[24] that the development of the first thermodynamical notions had begun already in the Old Stoa. Cicero quotes Cleanthes' doctrine which describes the function of heat in organic nature as a special case of thermic processes: "It is a law of Nature that all things capable of nurture and growth contain within them a supply of heat, without which their nurture and growth would not be possible; for everything of a hot, fiery nature supplies its own source of motion and activity; but that which is nourished and grows possesses a definite and uniform motion. . . . From this it must be inferred that this element of heat possesses in itself a vital force that pervades the whole world." Here the active character of heat as such and its connection with dynamic phenomena is expressed very clearly. Cicero goes on to illustrate thermic effects by various examples, such as generation of heat by friction, and the role played by heat in melting, evaporation, etc.

The greater emphasis placed on the active nature of air and fire during the period of the Old Stoa brought into greater prominence the contrast between the active elements and the passive ones—water and earth. The mixture of air and fire which was identified in Stoic physics with the pneuma thus became the active agent *par excellence* in their cosmos. Pneuma or one of its components was also defined by the collective "pneuma-like matter", and the Stoics attributed to them the property of coherence in the twofold sense of being cohesive and making cohesive,[25] whereas the passive elements ("hyle-like matter") were denied any such faculty. Without the active interference of air or fire or the two mixed together, the passive elements would disintegrate as they themselves do not possess the "cohesive force".[26] Coherence thus appears here distinctly as a force, and the geometrical continuum of Aristotle is in this way transformed into a dynamic one. It should be mentioned, however, that Aristotle has occasionally used the term "cohesive" in a structural

---

[23] Diog. Laert., VII, 156: φύσιν εἶναι πῦρ τεχνικὸν ὁδῷ βαδίζον εἰς γένεσιν.
[24] Cicero, *De nat. deor.*, II, 23-28.
[25] Galen, *De multitud.*, 3 (Arnim, II, 439 and 440).
[26] συνεκτικὴ δύναμις.

sense; e.g. in his analysis of the concept of unity[27]: continuity is achieved by a binding agent and not by geometrical contact alone, as for instance a faggot which is made one by its string, and pieces of wood by glue. This is a static analogy, but immediately afterwards Aristotle comes very close to a dynamic conception when he calls a thing continuous "whose motion is essentially one and cannot be otherwise". But the Stoics undoubtedly identified continuous with cohesive and thus completed the transformation of the geometrical concept into a physical one.

The most striking proof of this is the specifically physical property ascribed by the Stoics to the pneuma, or the "pneuma-like matter" in general: tension (*tonos*). On the one hand this latter term was also applied to the living body as shown by a quotation from Cleanthes given by Plutarch, where the vital force of the soul is defined as follows: "*Tonos* is the heat of fire which, if originating in the soul in sufficient measure to accomplish the task, is called strength and force."[28] Galen uses the expression "vital tension",[29] and according to another source, the Stoics saw the cause of sleep in a relaxation of the sensory tension.[30] On the other hand, several sources such as Plutarch[31] and Alexander[32] quote the Stoic theory of the function of *tonos* in inorganic matter. These quotations and the comments which accompany them show the gist of this theory: the tension innate in air and fire endows these elements with cohesion which is acquired by water and earth only through their admixture. Pneuma especially, which according to Galen was elevated by the Stoics to a "fifth quality",[33] possesses this tensional power as its most conspicuous property. As a matter of fact, since pneuma pervades the whole universe, the pneuma-like *tonos* makes the cosmos into a single cohesive unit[34] and thus the pneuma becomes the first version of the aether with all the characteristic functions ascribed to it from the seventeenth century on. We shall return later to this close analogy between pneuma and aether.

[27] Arist.,*Metaph.*, 1015 b 36 f.
[28] Plut., *De Stoic. repugn.*, 1034 d.
[29] Galen, *De locis affect.*, V, I (Arnim, II, 876).
[30] Diog. Laert., VII, 158.
[31] Plut., *De comm. not.*, 1085 d.
[32] Alex. Aphr., *De mixt.*, 223, 34.
[33] Galen, *Introd. s. medic.*, 9 (Arnim, II, 416).
[34] Clemens Alex., *Stromat.*, V, 8 (Arnim, II, 447).

The singular significance attached in Stoic physics to air and fire by their being components of the pneuma finds its expression in still another shift in their character, parallel to that which occurred with the "thermal neutralization" of air mentioned above. In pre-Stoic physics air and fire were generally regarded as light and thus rising, water and earth as heavy and thus descending. Aristotle, however, declared that air was both light and heavy at the same time,[35] but he said this in a very specific context. He wanted to describe the role of air in his dynamics of "forced" motion: i.e. in movements which are not "natural" the body is pushed up and down by air.

A survey of the sources on the Stoic view of the structure of air and fire[36] shows at first sight a certain semantic ambiguity. Sometimes these elements are denoted as "light" and sometimes as "not heavy". However, two of these sources which are fairly reliable are at the same time quite definite on this point. One of them, Plutarch, gives two quotations from the second book of Chrysippos' work *On Motion* and from his *Physical Arts*, both of them referring to air and fire as "non-heavy"[37]: "Fire is non-heavy and rises, and the same applies to air, so that water has to be assigned rather to earth and air to fire." "Air in itself possesses neither gravity nor lightness." The other source is Arios Didymos, who refers to the Stoic and especially to Zeno's cosmological conceptions.[38] The stability of the cosmos is secured by the tendency of all its parts towards the centre, and particularly of those which possess gravity. "But not all bodies have gravity, for air and fire are non-heavy. These spread over the centre of the whole sphere of the cosmos and bring about the union with its periphery. By their nature they frequent the upper parts because they have no share of gravity. He (Zeno) says likewise that the cosmos has no gravity because it is all composed of heavy elements and non-heavy elements." It is significant that the term "levity" or the attribute "light" has been avoided throughout the passage. It is also not said of air and fire that they are "rising" but that they "frequent the upper parts". The whole emphasis in this

---

[35] Arist., *De caelo*, 301 b 23.

[36] Cf. the items ἀήρ and πῦρ in the index of Arnim's *Stoicorum veterum fragmenta* (vol. IV).

[37] Plut., *De Stoic. repugn.*, 1053 e.

[38] Stob., *Eclog.*, I, 166, 4 (Arnim, I, 99).

text is on the gravitationally neutral character of air and fire who both participate in the cosmic tendency towards the centre, and at the same time stretch from there to the extreme regions, and thus contribute to the communication between all the parts of the universe. Obviously some of the essential characteristics of the pneuma are attributed here to its two components air and fire, and it is in this sense that we have also to interpret the last sentence of the text quoted. If the cosmos had gravity, the pronounced pull of the heavy elements towards its centre would make it collapse. But it is kept extended because of the admixture of the non-heavy elements which spread everywhere from the centre to the outer regions of the cosmic sphere.

2. *The Physical State of a Body*

Cohesion of matter is not the only effect produced by the tension of the pneuma. The latter has a twofold function: besides being a binding force, it is an agent which generates all the physical qualities of matter. From the point of view of the history of ideas we have to see in Anaximenes the father of this notion, in view of the function he assigned to air, out of which all matter originated by condensation and rarefaction. The next step in the direction of the Stoic conception was made by Diogenes of Apollonia who took up Anaximenes' basic idea and combined it with elements from Empodocles' and Anaxagoras' teachings. We shall be returning to Diogenes very soon. By their conception of the pneuma as the generator of physical qualities the Stoics generalized their continuum theory into a field theory; the pneuma is the physical field which is the carrier of all specific properties of material bodies, and cohesion as such thus gets a more specific meaning by becoming *hexis*, the physical state of the body. The following quotation from a book by Chrysippos *On Physical States* is very instructive: "The physical states are nothing else but spirits, because the bodies are made cohesive by them. And the binding air is the cause for those bound into such a state being imbued with a certain property which is called hardness in iron, solidity in stone, brightness in silver."[39] And a little later he continues: "Matter, being inert by itself and sluggish, is the substratum of the properties, which are

[39] Plut., *De Stoic. repugn.*, 1053 f.

7

pneumata and air-like tensions giving definite form to those parts of matter in which they reside." This gives some idea of the central position in the Stoic theory of matter of *hexis*, which denotes the structure of inorganic matter in a similar way to which *physis* expresses organic structure and *psyche* the structure of the living being. Galen[40] and Philo[41] describe the analogy between these three notions and their delimitations. Philo calls hexis "a very strong bond", and in another context he characterizes it as "a bond not unbreakable but hard to dissolve".[42]

The structural concept of hexis, also defined as the "binding spirit" of a body,[43] represents the highest entity in the hierarchy of inorganic structures as conceived by the Stoics.[44] These entities are divided into discrete, contiguous and unified.[45] At the lowest level we have an assembly of bodies in a disordered state, such as a crowd which does not lend itself to numerical determination. The following level is also a discrete state, but here the elements are arranged in an order which allows for numerical determination, such as a choir or an army in formation. This is a "denumerable" entity.[46] Contiguous structures are composed of conjoined elements, like the links of a chain or the planks of a ship or the stones of a house.[47] What the discrete and contiguous structures have in common is that each of their elements can continue to exist even if the rest are destroyed,[48] which is characterized by a simple additive relationship between the elements. True, the whole structure is more than the sum of its units (except in the case of disordered discrete structures which are very much the same as Lucretius' *concilia* of atoms), and each of these structures is held together by forces which have to be overcome before the whole is reduced to the sum of its elements. But each of n given elements would not be affected by anything that happens to the $(n-1)$ others, in fact there is no "communication"[49] between them which can lead to such a result.

[40] Galen, loc. cit. (Arnim, II, 716).
[41] Philo, *Leg. alleg.*, II, 22 and *Quod deus sit immut.*, 35 (Arnim, II, 458).
[42] Philo, *De incorrupt. mundi*, 24 (Arnim, I, 106).
[43] Achill., *Isagoge*, 14 (Arnim, II, 368).
[44] Sext. Emp., *Adv. math.*, IX, 78; Plut., *Praec. conjug.*, 34: *De def. orac.*, 29; Achill., *Isagoge*, 14.
[45] διεστῶτα, συναπτόμενα, ἡνωμένα.
[46] ἀριθμῷ ληπτός.
[47] Seneca, *Natur. quaest.*, II, 2.
[48] Sext. Emp., loc. cit., 80.   [49] διάδοσις.

The situation is completely different in the case of unified structures, such as stone or wood or metal which are "ruled by a single state".[44] It is not an additive principle which defines the relation between the physical state of a body, *hexis*, and its elements. We must realize that the elements of *hexis* are not mere localized units but physical properties which interpenetrate and create a totality where each of them shares in the existence of the rest. In our modern terminology: all the qualities which define the physical state of a certain body—its mechanical, thermic, electric, optical properties—have their origin in common roots and are therefore interdependent and not additive. Every one of them is affected if all or some of the others change. The physical state is an organization of dynamic character, each of its elements subsisting only in co-existence with the rest, and not able to exist if the organization as a whole disintegrates. The Stoic term for this form of co-existence of the elements of the highest structure was "sympathy" (*sympatheia*), and it is again significant that analogies from the living organism were given to exemplify this condition: when a finger is cut, the whole body shares in its condition.[48] It goes without saying that the living body was regarded similarly as a "united structure", as was shown, e.g. in the passage in Galen where he describes the faculties of the human body as structural elements of its physiology, extending throughout the whole body.[50]

The fundamental difference between the physical state which is the sum total of the natural properties of a body and structures of a lower order was apparently stressed very strongly in Stoic writings. Even so late a source as Simplicios raises this point in several passages, and one of them is especially illuminating.[51] It seems that Stoic terminology distinguished between qualities of the lower structures[52] which are produced to a greater or lesser degree by the co-operating effects of their units (e.g. the performance of a choir reached by the common efforts of each of its dancers individually), and the physical properties[53] which are immanent in the "unified" structures. Only these, the *hexeis*, are distinguished by a "pneuma-like unity" which is ruled by one law. The essential characteristic of the *hexeis*, missing in all the

---

[50] Galen, *De nat. facult.*, II, iii, 82.
[51] Simpl., *Categ.*, 214, 24 ff.
[52] ποιά.   [53] ποιότητες.

lower structures, is defined as the "coalescent and interlacing union"[54] of their properties. It is again significant that the term "coalescent", which originally was used mainly in biological contexts, is transferred by the Stoics to describe the highest order of inorganic structure.

The unmistakable connection between hexis and pneuma raises the question as to the mechanism by which different physical states were thought to arise out of the pneuma pervading the bodies. The very scanty remnants of Stoic physics do not permit us more than vague conjectures about the details of their theory on this topic. Some light is thrown on this problem in the passage in Galen already mentioned above[9]: "According to them (the Stoics), psyche is a sort of pneuma, as is also physis, but the pneuma of physis is more humid and colder, while that of psyche is drier and hotter. This pneuma is a kind of matter akin to psyche, and the specific quality of this matter is given by the proportions admixed of the airy and fiery substance." Here a principle of differentiation is stated for the world of animals and plants, derived from the basic assumption of the composite nature of the pneuma which allows for different ratios of mixture of its two components—a definite proportion of this mixture defining a definite quality.

One must emphasize here the unmistakable influence which the teachings of Diogenes of Apollonia had upon the Stoics in this respect. The fragments of Diogenes, preserved in quotations by Simplicios, clearly show how he had used Anaximenes' ideas to explain basic facts of anthropology.[55] Diogenes identifies soul and intelligence with air, and the variations in the former are explained by the difference in proportion of hot and cold, dry and wet in air. The differences in the degree of warmth of the soul in animals and in human beings is particularly stressed by Diogenes. The multitude of possible variations of the mixture is the reason for the observed diversity in form, way of life, and intelligence. The affinity between these ideas and the Stoic principle of differentiation seems to be rather striking; additional evidence for the influence of Diogenes' teachings on the Stoics will be brought later on.

We have seen already that hexis was regarded by the Stoics as

---

[54] συμφυὴς πρὸς ἀλλήλας ἕνωσις.
[55] Simpl., *Phys.*, 151, 28; see also Diels, 64 B 5.

a notion equivalent in the inorganic realm to that of physis and psyche in the organic, and we may therefore assume by analogy that here, too, the same principle of differentiation was used in Stoic physics. In other words, each of the physical properties of a body, according to this principle, would be defined by a specific sort of pneuma, characterized by a definite mixture of fire and air. All these different sorts of pneuma permeate the body and mix with each other without losing their identity, since the properties to which they are co-ordinated are well defined and stable in spite of the fact that by coalescing they form the specific nexus by which an inert lump of matter is endowed with hexis.[39]

3. *The Problem of Mixture*

We are thus led to the conclusion that the Stoic theory of hexis was based in a double sense on the process of mixture. On the one hand, mixture of pneuma with inert matter imbues the latter with physical properties, whereas pneuma itself, on the other hand, is a mixture of two components, fire and air. This easily explains the fact that the Stoics were occupied to such a large extent by the problem of mixture and that in several respects their theory of mixture goes beyond that of Aristotle.[56] The main sources are Alexander Aphrodisiensis[57, 58], Plutarch,[59] Plotin[60] and Arios Didymos.[61] Although they are of a fragmentary nature they give us some idea of how the Stoic view on mixture was heavily influenced by their extreme notions of continuity and their basic assumptions concerning the function of the pneuma. The passage in Aristotle just referred to already accounts very clearly for the difficulty with which a supporter of continuity is confronted with regard to this problem. If we assume that mixing of, say, two ingredients results in a mosaic-like composition of the components in the way that grains of barley and wheat are mixed with each other, then we can never claim that a mixture could be as homogeneous in all its parts as a pure substance. Let the particles of each component be as small as you like—the result

---

[56] Arist., *De gener. et corr.*, I, 10.
[57] Alex. Aphr., *De mixt.*
[58] Alex. Aphr., *Quaest. et solut.*, II, 12.
[59] Plut., *De comm. not.*, ch. 37.
[60] Plotin, *Ennead.*, II, vii.
[61] Stob., *Eclog.*, I, 153 (Arnim, II, 471).

would always be a composition (*synthesis*) which only macroscopically would give the appearance of a homogeneous mixture (*krasis, mixis*).

Aristotle states: "If mixing has taken place, the mixture ought to be uniform throughout and, just as any part of water is water, so it should be with a mixture."[62] For Aristotle the solution of the problem was given by his distinction between "actual" and "potential" existence: the components which are "capable of action and capable of being acted upon" combine to form a compound which is "actually" something different, while each component is still "potentially" what it was before the mixture. It is difficult to say what exactly Aristotle had in mind; was it something similar to a chemical compound in our terminology? In any case, his idea that mixture results in a kind of mutual assimilation if the components form a homogeneous whole led him seriously astray. If one component is predominant in bulk, Aristotle continues, then the mixture results in fact in a change of the weak component into the predominant one. "A drop of wine does not mix with ten thousand measures of water, for its form (*eidos*) is dissolved and changes so as to become part of the total volume of water."[63] Aristotle does not indicate at what point the ratio of the components results in a state of balance where the mixture still has properties depending on both the components, and from which point the weaker component loses its identity and is transformed into the stronger one. This instance serves only to emphasize how unsatisfactory from a physical point of view is Aristotle's theory of mixture.

As far as classification is concerned, the Stoic theory is much clearer. It distinguished between three types of mixture. One of them, mingling, or mechanical mixture, is identical with what Aristotle defines by "composition" (as in the case of the mixture of barley and wheat), and it applies essentially to bodies of a granular structure where a mosaic-like mixture results, each particle of one component being surrounded by particles of the other. The other extreme is fusion, which leads to the creation of a new substance whereby the individual properties of each of the components are lost, as in the case of drugs. This case corresponds to that which today would be called a chemical compound. Between these two types lies a third case of "mixture" proper

[62] Arist., loc. cit., 328 a 11.   [63] loc. cit., 328 a 27.

(*krasis* for liquids, *mixis* for non-liquids) which, from the Stoic point of view, represents the most important category of blending. Here a complete interpenetration of all the components takes place, and any volume of the mixture, down to the smallest parts, is jointly occupied by all the components in the same proportion, each component preserving its own properties under any circumstances, irrespective of the ratio of its share in the mixture. The properties are preserved in all cases where—as opposed to the case of fusion—the components can be separated out again from the mixture by physical contrivances. One special instance is mentioned by three sources,[61, 64, 65] namely the separation of wine and water which allegedly can be carried out by putting a sponge into the mixture. A plausible explanation of the nature of the described effect can be found in a passage of Alexander's polemics against the Stoics,[66] which reads: "The separation of the components becomes also evident by the difference of colours." The sponge apparently acts as an adsorbent, and by reducing the turbidity of the wine and decolourizing it, it gives the liquid squeezed from the sponge a more watery appearance and thus the illusion of separation is created.

On the whole, *krasis* would apply to what in the language of today is called a mixture in any of the three phases, or to a solution or a suspension. Of special interest to the Stoics were, of course, cases of extreme dilution or rarefaction, because they regarded the pneuma which permeates bulky matter without losing its properties as one of these cases. Chrysippos, obviously referring to the passage in Aristotle quoted above,[63] stresses this point with deliberate exaggeration: "There is nothing to prevent one drop of wine from mixing with the whole sea."[67] Other examples are given of extreme conditions of dilution or dispersion where minute fractions of a substance preserve their properties in a large bulk of another substance: frankincense burnt and rarefied in a large volume of air, and gold finely suspended in certain drugs.[68] In the latter case it is pointed out that this rarefaction could never be brought about by beating the pure metal. The Stoic view was that the medium of the more abundant

[64] Philo, *De confus. linguar.*, 184 (Arnim, II, 472).
[65] Alex. Aphr., *De mixt.*, 232, 1.
[66] loc. cit., 233, 6.
[67] Plut., loc. cit., 1078 e.
[68] Alex. Aphr., *De mixt.*, 217, 13 ff.

component actually assists both the expansion of the weaker one throughout such large volumes and its homogeneous occupation of the whole space offered to it.

It is very interesting to note how this picture of a homogeneous dilution, so trivial for us today, met with general opposition as far as one can judge from the extant writings of the Aristotelian commentators and other philosophers of late antiquity. One of the reasons for this was doubtless the failure to grasp the meaning of specific density, in spite of Archimedes' achievement which clearly had not succeeded in uprooting the old dialectic conception of the opposition of heavy-light, dry-wet, etc. While the Stoics realized very well that the same substance could exist in different densities according to varying physical conditions, this notion was rejected even by critics who on the whole were favourably inclined towards the Stoic theory. Plotin,[69] for instance, could not accept the enormous "stretching of masses"[70] which occurs when a small quantity of one liquid is mixed with a large quantity of another. According to him, this phenomenon could not be compared with the increase in volume taking place when water evaporates, as in the latter case there occurs a change in the physical state. Even this phenomenon, Plotin characteristically adds, is of a problematic nature, and it is not clear to him what actually happens when a certain quantity of water "becomes air", thus increasing its volume.

Plutarch, giving an account of the Stoic theory of mixture, also reveals in his criticism a complete lack of understanding of the nature of dilution. According to the Stoics, he says, one measure of wine mixed with two measures of water should give four measures of mixture, as the wine will extend into the whole volume occupied by the water.[71]

The main reason for the rejection of the Stoic theory of mixture, however, was the opposition to their hypothesis of "total mixture".[72] Their conception of a homogeneous distribution of the components throughout the mixture, "whereby there is no part of the mixed substance which does not participate in the mixture as a whole",[68] was of course a sound and logical consequence of their idea of continuity. Every phenomenological theory of continuity argues in the same way even today where

---

[69] Plotin, loc. cit., II, vii, 1, 52-2, 2.
[71] Plut., *De comm. not.*, 1078 a.

[70] ἔκτασις τῶν ὄγκων.
[72] κρᾶσις δι' ὅλων.

atomic structure is an indisputable fact, provided that the volumes which are taken into account are large compared with atomic dimensions. But the Stoics, taking a radical position with regard to continuity, conceived of mixture as a complete inter-penetration of the components which existed simultaneously in the given proportions down to the most minute elements of volume. This conception of total mixture was understood by them in the sense that every element of volume, however small, would be homogeneous with regard to the mixing of the components, and that from no point on would this homogeneity dissolve itself into a mosaic-like structure with bits of the components lying side by side. Alexander Aphrodisiensis and all the later critics—with one exception—saw in this notion of total mixture an infringement upon the principle that one body cannot occupy the place occupied by another. Their view was[58] that complete interpenetration of bodies must necessarily lead to the volume of the mixture not being the sum of the volumes of the components, but instead it would remain constant as compared with the volume of the largest of the components.[73] Plotin[74] gives a two-dimensional picture in explanation: Total mixture would be comparable not to the juxtaposition of two (material) lines drawn side by side—which would mean an increase in area—but to the superposition of the two lines whereby no increase occurs. In view of the obvious fact that generally a mixture occupies a larger volume than each of its components, the Stoic theory, as inter-preted by its critics, was generally rejected. Plotin alone, in his endeavour to give the Stoic ideas a fair trial, attempts to reconcile their views with the facts.[75] Volume, he says, is one of the many qualities of a body. According to the Stoics, the qualities of the components are preserved in the mixture; in fact, the properties of the latter result from the mixing of those of its components. In the same way, he concludes, the mixture will possess a volume which is exactly or nearly the sum of the volumes of the components—thus resulting in an increase of volume.

We must look at the Stoic theory of total mixture in the light

---

[73] Cf. Simp., *Phys.*, 530, 29: "A ladle mixed with another one in total mixture shall give again a ladleful."
[74] Plotin, loc. cit., II, vii, 1, 45.
[75] Plotin, loc. cit., II, vii, 1, 35.

of the main purpose it was supposed to serve—to give a firm foundation for its conception of hexis. We must remember that pneuma was thought to be an extremely rarefied substance and that its total interpenetration of matter would thus create a case analogous to that of the mixture of a small drop of wine with a large measure of water, i.e. of one component being negligible in bulk as compared with the other. Here again the starting-point for the argument was the organic world. Soul (*psyche*), the hexis of the living body, was itself corporeal according to the Stoics, who included it in the group of things which are capable of acting and being acted upon. Mutual interpenetration of soul and body, of physis and plants, of hexis and inorganic matter, have all one common feature—the total mixture of a very tenuous and rare component, the pneuma, with a much bulkier one.[76], [77] The common denominator in all these cases are the physical qualities or properties which themselves are nothing less than pneumata mixed in definite proportions. And as we have seen, qualities are bodies, like the soul itself, and therefore their mixture with matter—organic and inorganic alike—must be of the nature of a total mixture.[78] The later critics of this theory, especially the Neo-Platonists, reversed the argument: the soul must interpenetrate the body totally; therefore it cannot itself be a body.[79], [80] Galen touches upon the subject in another context[81] which proves that the conception of total mixture of *qualities* goes back to Hippocrates.

At about the same time and possibly slightly earlier (middle of the fifth century B.C.) Anaxagoras developed his theory of seeds which in a certain way can also be regarded as a precursor of the total mixture doctrine of the Stoics. The infinitely divisible seeds of each form of matter contain something of all the opposite qualities, though in different proportions. The Stoics were the first to ascribe the faculty of mixing totally to *substance*, or rather to the corporeal nature of qualities by identifying them with the pneuma in different states of composition. However, the most significant difference between Anaxagoras' theory and that of

---

[76] Galen, *Comm. 5 in Hippocr. epid.* 6 (Arnim, II, 715).
[77] Alex. Aphr., *De anima*, 26, 10.
[78] Simpl., *Phys.*, 530, 9.
[79] Plotin, *Ennead.*, IV, vii, 10.
[80] Chalcidius, *In Timaeum*, 221 (Arnim, II, 796).
[81] Galen, *De nat. facult.*, I, ii, 5.

the Stoics is to be found in the much wider conception of the latter, where the idea of interpenetration is linked up with the tensional qualities of the pneuma and thus, as we shall see later, with the notion of the field of force.

4. *The Four Categories*

Before concluding the chapter dealing with the basic Stoic conceptions of the continuous character of the world, we have to consider a methodological principle stated by them which has a direct bearing on the framework of their physical theory, and was possibly developed first to fit the conceptual frame of physics and later generalized as a directive for other fields of scientific and philosophical thought. This methodological principle is known as the Stoic doctrine of categories, but has hardly anything in common with the Aristotelian classification of concepts known by the same name. Aristotle's ten categories[82] are an attempt —in accordance with a principle not known to us—to compile a list of concepts such that every word of non-compounded meaning can be shown to belong to one of them. This list thus forms a horizontal classification. In other words—the various points of view from which one can look at objects are regarded by Aristotle as a group of co-ordinated notions not bound together by a higher notion embracing all of them.

The four Stoic categories, on the other hand, are a vertical classification according to which every object is determined by four successive steps of increasing specification such that every category includes the preceding one. They thus represent a kind of methodological guide for the complete ontological definition of an object. There is a highest notion, comprising all the four categories, *the something* followed by the four categories *substratum, quality, state* and *relative state.*[83]

The ancient sources, especially those dealing with Stoic logic, give only fragmentary and sometimes not very clear expositions of the four categories[84] and probably it has rightly been pointed

[82] Substance, quantity, quality, relation, place, time, position, state, action, affection.
[83] Simpl., *Categ.*, 66, 32 f.; Plotin, loc. cit., VI, i, 25: ὑποκείμενον, ποιόν, πως ἔχον, πρός τί πως ἔχον.
[84] Cf. B. Mates, *Stoic Logic*, p. 18.

out that the Stoic categories belong rather to their physics.[85] Indeed, it seems doubtful whether metaphysical and ethical subjects have been completely analysed by means of the categories. An attempt has been made to prove this from the writings of Epictetos, Marcus Aurelius and others.[86] Neither can a satisfactory proof be given in the case of physics, but sufficient evidence is supplied by the extant sources to make such an assumption a plausible one.

If we look at the conceptual structure of the physical continuum as we have come to know it in the preceding pages, we perceive a striking analogy with the set-up of the Stoic categories. Shapeless and passive matter is the primary substratum of the cosmos and as such without any qualities.[87] It is the all-pervading pneuma which, by totally mixing with matter, imbues it with all its qualities,[39] and thus represents the second category. The third category, the state, is given by the fact that to each specific quality of the body there is attributed a specific mixture of the pneuma defined by a certain proportion of its two components, air and fire. The sum total of all pneumata permeating the body then defines its physical state. There is a lack of consistency in the extant sources with regard to terminology. Thus we find no direct reference to hexis as a physical state, but Alexander once mentions the "physical property of a body", which is much the same as hexis, and calls it "pneuma in a certain state".[88] Hexis, however, is mentioned expressly in connection with the fourth category, and the passage in Simplicios referring to it[89] is of great significance for the understanding of the whole subject. Simplicios informs us of a subdivision of the fourth category: the Stoics distinguished between two kinds of relations, the *relative state*[90] and the *relative*[91]. The first denoted a state defined by that of another thing outside it, such as the relation father-son or left- and right-hand neighbour. The *relative* referred to things capable of change (the examples given are sweet and bitter), whereby

[85] J. M. Bochenski, *Ancient Formal Logic* (North-Holland Publish. Comp. 1951), p. 87.

[86] P. de Lacy, *Trans. Amer. Phil. Assoc.*, Vol. 76, p. 246 ff. (1945).

[87] Diog. Laert., VII, 137.

[88] Alex. Aphr., *Topica*, 360, 10.

[89] Simpl., *Categ.*, 165, 32-166, 29.

[90] πρός τί πως ἔχον.

[91] πρός τι.

the relation is given through comparison of two states of this change (e.g. two degrees of sweetness). Simplicios quotes hexis as another example of the relative which throws into relief the dynamic notion of Stoic physics, because hexis is the key term of the physical continuum which embraces an infinity of different states. Each of these states can evolve from another by a continuous transition produced through the "change of the former quality",[92] a change which corresponds to that of the spectrum of all pneuma tensions permeating the body involved.

Simplicios further reports the discussions on how far the two subdivisions of the fourth category are interrelated, and whether the first implies the second or vice versa, or whether they are independent of each other.[93] We shall not touch upon this question here, but it is worth while to note that on this point the views of the earlier Stoics differ from those of the later. However, it is evident that, as far as the description of the physical world is concerned, the subcategory *relative state* completes the set-up described above. The relative state determines the relation between the physical states of two different bodies and, taking into account the fact that each of these bodies can itself undergo changes corresponding to a continuum of states, it is obvious that all the possible combinations resulting from this fact together form the totality of physical occurrences.

Admittedly the above account is not so much strict evidence as a mere indication of a methodological connection between Stoic physics and their categories. It is, however, highly significant how well the Stoic categories correspond to the methodological schema of Newtonian dynamics. Here we have a system of bodies forming the *substratum*; the *quality* is given by the various traits exhibited by these bodies, such as spatial distribution, mass, velocity, etc. If the specific data of these traits are known for a given moment, the *state* of the system is hereby defined. This state can undergo changes with time and the new state is describable in terms of the former which defines the *relative*. The relation, finally, between the states of two different systems at any moment determines the second subdivision of the fourth category, the *relative state*. This again goes to show that in all probability

---

[92] Galen, *Method. med.*, I, 6 (Arnim, II, 494).
[93] Simpl., loc. cit., 167, 20-36.

the four categories were constructed in the first instance to fit methodologically the conceptual set-up of a physical theory, and moreover it hints at the structural similarity between Stoic continuum physics and the classical mechanics of our time, as will be elucidated further in the following chapters.

II

PNEUMA AND FORCE

―――――――

1. *Dynamics of the Pneuma*

THE specific Stoic attitude which attributes corporeality to everything capable of acting and being acted upon, including even the soul, had a decisive influence on the development of the dynamic aspect of their notion of continuity. The conception of the soul as a motive force can be traced back right to the beginning of Greek science, as pointed out by Aristotle in a well-known passage of *De anima*[1]: Thales seems to have attributed a soul to the magnet in order to explain its action on iron; Alcmaeon held that the soul is immortal because of its ceaseless movement, "for all the things divine, moon, sun, the planets, and the whole heavens, are in perpetual movement". The sentence alluding to the divinity of the heavenly bodies is apt to shift the emphasis from the central issue. If Alcmaeon's view has been correctly rendered by Aristotle, it was not solely a question of the soul being a motive force, but of the soul itself being in perpetual motion (and being in turn the source of other movements). It was precisely this conception of the soul as something in motion, combined with the notion of its corporeal nature, that led to the Stoic theory of the dynamics of the pneuma. The essence of this theory which we are about to discuss now was to conceive of hexis—the synthesis of pneumata permeating the body—not as a static

―――――――

[1] Arist., *De anima*, 405 a 20 ff.

phenomenon but as a dynamic process prevailing within a continuous medium and "of a corporeal nature".[2]

The biological approach to nature characteristic of Greek science in general often makes it imperative to begin with the analysis of physical concepts in the field of biology. In order to understand the Stoic dynamics of continuous media, we have to start from their theory of sense perception.[3] An essential part of this theory is the hypothesis that there exists a centre of perception and consciousness, the "ruling part of the soul" (*hegemonikon*) which is located in a defined part of the body (most of the Stoics identified it with the heart), and which is in control of the five sensory organs, the generative part of the body, and the faculty of speech. Every stimulus from outside is conducted from the specific sense organ excited by it to the hegemonikon, which centralizes and co-ordinates the various impressions, elevates them into consciousness and then releases the impulse reacting to the sensation. The vital function of the hegemonikon as the central seat of consciousness, unifying all the activities of the soul and maintaining and regulating its contact with the external world, clearly defines a dual direction of communication: from the centre of the body to the various organs on its surface, and vice versa. It is through the incessant movement of pneuma to and fro between the hegemonikon and the surface of the body that this two-way communication is established,[4] and Alcmaeon's "perpetual motion" of the soul thus acquires a well-defined meaning in the framework of the Stoic picture of the connection between mind and sensation.

What kind of motion did the Stoics have in mind when they talked about movement of pneuma within the body? Modern physics distinguishes between two types of motion: particle movement, i.e. transport of matter, and wave motion, i.e. propagation of a state. There can be little doubt as to the Stoic attitude in this respect; their conception of continuity and the idea of tension inherent in the pneuma make it highly probable that they visualized movement of pneuma as something akin to the second type of motion, the expansion of a disturbance in an

[2] Stob., I, 371, 22 (Arnim, II, 801).

[3] For details, cf. e.g. Pohlenz, *Die Stoa*, I, 87 f. ; II, 51 f.; and Brehier, *Chrysippe*, p. 164 ff.

[4] Philo, *Quaest. et solut. in Genesin*, II, 4 (Arnim, II, 802): "anima . . . quae in medio consistens ubique permeat usque at superficiem in medium vertitur."

elastic medium. Indeed, the Stoics were the first to advance the hypothesis of a circular or spherical propagation of a disturbance in air, and to use the analogy of water waves: "The Stoics say that the air is not composed of particles, but that it is a continuum which contains no empty space. If struck by a puff of breath it sets up circular waves which advance in a straight line to infinity, until all the surrounding air is affected, just as a pool is affected by a stone which strikes it. But whereas in this case the movement is circular, the air moves spherically."[5] And we are told similarly by another source about the Stoic theory of sound: "We hear when the air between the sonant body and the hearer is struck in spherical waves which impinge on the ears, just as the waves in a pool expand in circles when a stone is thrown into it."[6] There is no direct evidence of an analogous picture in the case of light and the theory of vision, but here again the Stoic theory emphasized the tension of the medium between the eye and the object. As is well known, the gist of the Stoic theory was that the "optical pneuma" emitted from the hegemonikon to the eye, excites[7] the air adjoining the pupil and that from there the object is contacted through the stressed air.[8, 9] The significant part of this hypothesis is the picture of a cone of air in a state of tension with its apex in the pupil of the eye and its base at the object. "The object seen is reported through stressed air, as if in contact by a stick."[10] The analogy of the stick is a clear indication of the rejection by the Stoics of the picture of emitted particles and shows the emphasis put on contiguity in the propagation of action. The notion of the space through which this action takes place as being in a state of tension, however, is the most striking proof of the new approach to the problem of transmission in a continuum.

Let us now return to the stationary movement of the pneuma in the human body by which communication is maintained between the hegemonikon and the sensory organs. In addition to the passages already quoted, two more analogies will furnish

---

[5] Aet., IV, 19, 4 (Arnim, II, 425).
[6] Diog. Laert., VII, 158.
[7] νύττειν.
[8] Cf. e.g. Aet., IV, 15, 3 (Arnim, II, 866).
[9] Alex. Aphr., *De anima*, 130, 15.
[10] Diog. Laert., VII, 157; cf. also Galen, *De Hipp. et Plat. plac.*, VII (Arnim, II, 865), and Gellius, *Noct. Att.*, V, 16, 2.

additional evidence for the assertion that the Stoics regarded the movement of the pneuma as some sort of propagation of a state. In one of them[11] the "seven parts of the soul" growing out of the hegemonikon and extending through the body are compared to the tentacles of an octopus—a picture somewhat similar to that of the stick quoted above. The arms of the octopus are in fact a prolongation of its central part and their stretched or unstretched position reflects and indicates the actual condition of the animal. It is interesting to follow the increasingly radical attitude which the Stoics took in elaborating their notion of the propagation of the pneuma in the body. Cleanthes defined walking (choosing a particular instance of an activity directed by the mind) as pneuma emitted from the hegemonikon to the feet, whereas his disciple Chrysippos maintained that it is the hegemonikon itself which reaches the feet.[12] It is obvious that Chrysippos was anxious to avoid the misconception of the convection of a spirit or its emission from the centre to the limb like a corpuscle or projectile. By insisting on the image of the hegemonikon itself extending to the boundaries he gave prominence to the idea of the propagation of an impulse through a medium in a state of tension. This idea is thrown into relief quite unambiguously in the second analogy,[13] quoted in the name of Chrysippos: "In the same way as a spider in the centre of the web holds in its feet all the beginnings of the threads, in order to feel by close contact if an insect strikes the web, and where, so does the ruling part of the soul, situated in the middle of the heart, check on the beginnings of the senses, in order to perceive their messages from close proximity." The term "by close contact" (de proximo) is a most striking expression of the Stoic principle of contiguous action which is here demonstrated by the propagation of an impact through the medium of the taut threads of the cobweb. It should be noted that in all probability the analogy of the spider in the centre of the cobweb was taken over by Chrysippos from a very similar one going back to Heracleitos.[14] The way in which Chrysippos has modified the version to suit his main purpose is highly significant. The gist of the Heracleitean fragment is, that when a wound is inflicted on

---

[11] Aet., IV, 21 (Arnim, II, 836).
[12] Seneca, *Epist.*, 113, 23 (Arnim, II, 836).
[13] Chalcid., *Ad Timaeum*, 220 (Arnim, II, 879).
[14] Diels, 22 B 67 a.

some part of the body, the soul immediately rushes to the sore place like the spider hurrying to the spot of the web damaged by the impact of the fly. The *tertium comparationis* is here the "strong and proper" (firme et proportionaliter) connection between soul and body on the one hand, and spider and web on the other. This is, of course, also the central idea in Chrysippos' version, but he discards completely the picture of locomotion from the centre to the afflicted point, because for him the very existence of the medium in a state of tension establishes the communication between the two points, the nature of this communication being characterized especially by the words "de proximo", which are an innovation by Chrysippos. The special subject mentioned in Heracleitos' fragment—the physiology of pain—was dealt with by the Stoics in accordance with the theory outlined above: pain is propagated from the afflicted part of the body in successive communication by the pneuma till it reaches the hegemonikon, and there it enters into consciousness and is felt by the soul as a whole.[15]

The Stoic theory of perception contains further evidence of the dynamic character ascribed by the Stoics to the pneuma. There was a difference of opinion between Chrysippos and his predecessors as to the mechanism of presentation (*phantasia*), i.e. the way in which perception takes place in the hegemonikon. Cleanthes' definition, apparently taken over from Zeno, was that "presentation is an impression (*typosis*) on the soul[16] or on the hegemonikon".[17] We are told that "impression" was taken by Cleanthes in the literal sense—"as involving depression and protrusion, just as does the impression made in wax by signet-rings", a notion which was strongly rejected by Chrysippos. He argued that impressions of different sensations would destroy each other and that especially the storage of presentations by memory would be impossible, as the latest impression would obscure the preceding one, in precisely the same way as that of a second seal obliterates the impression of the former.[18] Instead of this, Chrysippos defines presentation as "modification (*heteroiosis*) of the soul",[16] and, presumably out of reverence for the founder

[15] Plotin, *Ennead.*, IV, vii, 7.

[16] Sext. Emp., *Adv. math.*, VII, 228. Cf. Democritos' theory of vision and Theophrastos' criticism, Theophr., *De sensu*, 50-51.

[17] Sext. Emp., *Pyrrh. hyp.*, II, 70.

[18] Sext. Emp., *Adv. math.*, VII, 372.

of the School, maintains that Zeno had used *typosis* in the sense of *heteroiosis*. "It is by no means absurd", he says, "that the same body be submitted at one and the same time to a very large number of modifications. In the same way as air, when many sounds are uttered, is submitted to innumerable and different strokes and holds at once many modifications, so the hegemonikon undergoes an equivalent experience when presentations are formed by it in various ways."[16] The parallel drawn here with acoustic phenomena brings out clearly the dynamic notion of the pneuma. Sounds co-exist and do not obliterate each other, because they are dynamic processes in the air, of which each of them is a certain modification. Such modifications can undergo super-position without losing their identity, whereas a superposition of static states, like Cleanthes' impression, does away with each of them. We have thus to interpret Chrysippos' "modifications of the soul" as different dynamic states of the pneuma, existing by virtue of the latter's inherent tension and admitting of super-position which preserves each of them individually. It is some-what similar to total mixture where every component retains its special properties and can be separated out again from the mix-ture. However, the superposition of the modifications of the pneuma is of a specifically dynamic character since each modifica-tion is given by a definite movement of the pneuma. If we bear in mind that the Stoics had a clear conception of the wave character of sound, the analogy quoted might easily indicate that they already had an inkling of the principle of the superposition of waves.

The use of the term *heteroiosis* in a sense similar to that applied by Chrysippos can be traced back to Diogenes of Apollonia, a fact which constitutes one more proof of the influence of this younger contemporary of Anaxagoras on the Stoic theory of pneuma as mentioned above (p. 10). The basis of Diogenes' philosophy of nature is the monism of the Milesian School to which he gives the generalized formulation "all existing things are modifications of the same thing".[19] In following Anaximenes and putting the emphasis on air, especially for the purpose of his physiological theories, Diogenes amplifies the classical explanation that every-thing is transformation of air through rarefaction or condensation. In addition to heating and cooling (which of course are identical

[19] Diels, 64 B 2.

with rarefaction and condensation) he mentions as instances of the many possible modifications of air its getting drier and moister and steadier and moving faster.[20] Of these modifications, by which the diversities of the forms of life are explained, the latter ones are of the most significance for our subject. Diogenes was the first to attribute physical or physiological changes to modifications in the motion of the air. A few decades later, Archytas used the term *oxys* (fast) in contradistinction to *barys* (slow) in his theory of sound to explain the generation of high-pitched and low-pitched notes by quick and slow motions.[21] The analogy of sounds used by Chrysippos when explaining the modifications of the soul makes it not improbable that for him *heteroiosis* was in the first place a modification in the state of motion of the pneuma, preferably a motion arising out of and connected with its tensional properties. For lack of direct evidence we have to content ourselves with this attempt at reconstructing the meaning given to "modification" by Chrysippos. The ironical allusion of Sextus to "the strange 'modifications' they talk about"[17] need in no way reflect on Chrysippos' ability to define a scientific term properly, but might rather be another of the numerous instances of the lack of understanding (in antiquity as well as in modern times) for the scientific way of thought of the Stoics, who in many respects were ahead of their time.

We must now briefly give our attention to another aspect of the Stoic theory of vision. Here we are confronted with a mode of Greek scientific thought basically different from our modern way of describing the world around us. Whereas modern science strives to extend the mathematical picture of the physical universe into the realm of organic nature and of man, the opposite tendency prevailed in Greek science. There the concepts and pictures of biology were projected into the realm of the physical world, thus transforming the cosmos into a living organism. This way of seeing the world as an extrapolation of man is already thrown into relief by the basic Stoic assumption that objects are not seen by light coming from them and impinging on the eye, but by a reverse process. Actually for the Stoics the air between eye and object continues the process going on inside the human body between the hegemonikon and the eye, and the sense perception of man is followed up by that of the air which, by "seeing"

[20] loc. cit., 64 B 5.   [21] loc. cit., 47 B 1.

the object, completes the bridge between the centre of sensation and the object. According to this hypothesis, the faculty of perception is given to the air by the light: "Something similar happens to the air surrounding us. When illuminated by the sun it becomes an organ of vision precisely as the pneuma arriving (in the eye) from the brain, but before the illumination occurs which produces a modification through the incidence of the sun's rays the air cannot become such an affected organ."[22]

Galen who gives this exposition of Poseidonios' theory[23] draws a parallel between the "affection" or "modification"[24] of the nerves leading to the hegemonikon and that of the air leading to the eye. In the same way as the optical pneuma makes the nerves part of the hegemonikon and thus enables them to communicate their impressions, so the illuminated air becomes pneuma-like and is able to perceive. The version given by Poseidonios is confirmed by Cicero's formula: "The air itself sees together with us and hears together with us" (ipseque aer nobiscum videt nobiscum audet).[25]

The principle underlying this theory was that of the unifying power of equality, adopted by many schools of thought in Greece throughout the ages. "Things of the same kind get to know each other best" is a maxim used also by Democritos in his theory of vision.[26] By Alexander Aphrodisiensis we are informed as to the mechanism by which the Stoic theory explained the "modification" of the air: "The illuminated air becomes more powerful because of the mixture and can propagate the sensation through pressure, whereas dark air is slack and cannot be stressed by vision."[27] It is the greater tenuity of the air through its mixture with the fire-like light that gives it the properties of the pneuma and especially enables it to get into a state of tension. Alexander remarks that it would be more plausible to assume that dark air has greater "optical conductivity" because "it is denser than illuminated air". This shows that he fails to understand the central Stoic conception of a field of force. On the other hand, we do not know how the Stoics explained by their theory other arguments mentioned by Alexander, such as the argument that

[22] Galen, *De Hipp. et Plat. plac.*, VII (p. 641, Mueller).
[23] Cf. K. Reinhardt, *Kosmos und Sympathie* (Muenchen, 1926), p. 188 ff.
[24] ἀλλοίωσις.
[25] Cicero, *De nat. deor.*, II, 83.
[26] Theophr., *De sensu*, 50.   [27] Alex. Aphr., *De anima*, 131, 32.

one can observe in the dark an illuminated object in the distance.

The state of tension of the transmitting medium forms also the essential part of the picture in the mechanistic explanation going back to Chrysippos which was mentioned before (p. 23). The purely mechanical analogy of the transmission of light ("reported as by a stick") probably did not satisfy Poseidonios, and his criticism might well have been similar to that found in the passage in Alexander just quoted[28]: mechanical transmission could only communicate properties such as hardness or smoothness and not the typical optical ones such as colours.

2. *Tensional Motion*

The movement of the pneuma between the central seat of sensation and the peripheral sensory organs, akin to the propagation of a state in an elastic medium, was called by the Stoics "tensional motion" (*tonike kinesis*).[29] It is worth discussing this concept in greater detail, as it was not restricted to the domain of perception but found much wider application, and indeed represents an important clue for the understanding of the Stoic notion of continuity. Philo, using the term metaphorically to describe the propagation of the word of the Lord, posits *tonike kinesis* against translatory motion[30] and defines the latter as a motion "whereby one place is occupied and the other abandoned".[31] This definition clearly implies that tensional motion is unlike the change of place occurring when a body moves, i.e. unlike the translatory movement of matter, implying in fact that it is propagation in a continuous medium of a state of tension. Great as the temptation might be to identify *tonike kinesis* with wave propagation, we must refrain from doing so for lack of direct evidence. Nevertheless, this term borrowed from modern physics certainly describes the meaning of *tonike kinesis* more accurately than the usual interpretation as "pneuma currents", which does not take into account the *pneumatikos tonos* (ch. I, n. 34), the tension of the pneuma. The verbs usually associated with the movement of the pneuma are "pervade"[32] (which allows of any

[28] Alex. Aphr., loc. cit., 131, 22.
[29] Cf. Nemesios, *De nat. hom.*, ch. 2 (Arnim, II, 451).
[30] μεταβατικῶς κινεῖσθαι.
[31] Philo, *De sacrif. Abel et Cain*, 68 (Arnim, II, 453).
[32] διήκειν, see Alex. Aphr., *De mixt.*, 216, 15.

interpretation) and "extend" or "stretch",[33] which decidedly puts the emphasis on *tonos*. Each of the two opposite senses of the tensional motion has a specific function, as we have noted already. The motion towards the centre, in co-ordinating all the different sensations, produces the unity of consciousness, whereas the motion towards the periphery, maintaining the contact with the outside world, differentiates between the various qualities of sensation on the level of consciousness (cf. also ch. I, n. 31). This rapid succession of outward and inward motions, which follow on steadily as long as the body lives, can be seen as one single motion resulting from the superposition and merging of both movements into one. We find indeed the *tonike kinesis* sometimes defined as "simultaneous motion in opposite directions",[34] or "simultaneously moving inwards and outwards".[22] However, in some cases the alternation of the two motions is emphasized especially, as for instance by Galen, who, following in the footsteps of the Stoics, describes the "innate heat" of bodies and plants as "in perpetual motion, not moving inwards alone, but one movement constantly succeeding the other".[35] Galen draws an interesting parallel between the well-known Heracleitean picture of the eternal upward and downward motion which maintains the life of the cosmos and the inward and outward movement which represents a balance of the hot (outward) and the cold (inward). Most probably this parallel shows the influence of Cleanthes' teachings who regarded the sun as the hegemonikon of the cosmos.[36] In the same sense, tensional motion is characterized by Simplicios as "rarefying and condensing motion"[37] whereby the first—analogous to the upward movement of Heracleitos—is to be identified with the outward motion, and the latter with the inward one.

The whole conception of unity created in the living body through the bond of the tensional motion was transferred by the Stoics in a generalized form to physis and also to our main topic of interest, to hexis, the physical structure of inorganic matter. It has already been pointed out (p. 8) that the Stoics conceived physical bodies (usually denoted by "stone and timber") as held

[33] διατείνειν, see Stobaios, *Eclog.* I, 368 (Arnim II, 826).
[34] Alex. Aphr., *De mixt.* 224, 24: εἰς τὸ ἐναντίον ἅμα κίνησις.
[35] Galen, *De tremore*, 6 (Arnim II, 446).
[36] Euseb., *Praep. evang.*, XV, 15, 7.
[37] Simpl., *Categ.*, 269, 14.

together by pneuma in a state of motion, and that all the various pneumata permeating it constitute hexis, the sum of its physical properties. The movements of the pneuma are nothing else but tensional motions. Stobaios[38] speaks of "the pneuma moving by itself to and fro, or moving forward and backward". Philo,[39] who calls hexis a "very strong bond", defines it as pneuma "turning back the way it comes"[40] and amplifies this definition with the explanation: "It begins in the centre of the body and extends outwards to its boundaries, and after touching the outermost surface it turns back till it arrives at the same place from which it started." According to him, this continuous double course of the hexis (which he compares to the course of horse-races) is a never decaying one. It should be noted here that as in Philo the expression "bond" (*desmos*) is used by Alexander Aphrodisiensis[41] when he talks of pneuma as the cause of the unification of matter. Sextus[42] emphasizes that hexis controls mainly the conditions of the bodies "related to expansion and contraction", but it is clear from the context that what he has in mind are the specific physical states of bodies.

In spite of the scarcity of the extant sources which are relevant, the picture we get of the Stoic notion of hexis is a fairly clear one. Physical bodies have coherence and definite properties by virtue of the everlasting movement of a very tenuous and elastic medium pervading them. Since matter is conceived as strictly continuous, the medium performs its tensional fluctuating motions within matter itself, being united with it in total mixture. The dynamic concept of hexis by which the physical state of a body is defined is thus akin to what we would call today a field of force. The Stoic notion of continuity, applied to the phenomena of the physical world, has therefore led, by the intrinsic logic of scientific thought, to the concept of forces acting in accordance with the principle of continuity, these forces being the cause of the cohesion of matter as well as of its specific physical qualities. If we exclude Empedocles, who introduced the idea of two opposing forces into his cosmology in order to explain the dynamic equilibrium prevailing in the cosmos, it was the Stoic

---

[38] Stob., *Eclog.*, I, 153, 24 (Arnim, II, 471).
[39] Philo, *Quod deus sit immut.*, 35 (Arnim, II, 458).
[40] ἀναστρέφον ἐφ' ἑαυτό.
[41] Alex. Aphr., *De mixt.*, 223, 16.
[42] Sext. Emp., *Adv. math.*, IX, 82.

School who first conceived forces of a well-defined nature as being active in matter and giving it a definite shape. The ingenious picture of tensional motion by which the Stoics gave expression to the continuous as well as dynamic character of this force, at the same time confining its field of action to a body which as a whole might well be at rest, is similar to that of a standing wave or stationary vibration in modern physics. The problem of describing motion within a given volume which is macroscopically in a state of rest was also solved by the Greek atomists by the picture of random motion of the atoms of a body, the vector sum of their velocities being zero. Lucretius likens it to a flock of sheep which as a whole appears at rest from afar whereas the members of the flock are in an unceasing disordered movement.[43] One should bear in mind, however, that the atomic school denied the existence of forces between the atoms, restricting any mutual influence to impacts. The Stoics were the first to postulate continuous forces between parts of matter by introducing their dynamic theory of the physical state.

The influence of the Stoic theory of pneuma and tensional motion on the scientific thought of later antiquity is reflected equally in applications of these notions in various scientific writings and in the strong and sometimes violent reactions of opposing schools, especially the Peripatetics. Of special interest is the use Galen makes of the analogy of *tonike kinesis* in his study of muscular motion.[44] According to him, muscles of an extended arm under strain, being in a state of excitation (which implies movement) and yet at rest as a whole, offer a picture of tensional motion. It is not surprising that Galen, as the foremost authority of his age on muscles, veins and arteries, and their motions and pulsations, and as a firm adherer to the continuum theory of the Stoics, should have introduced the concept of a stationary movement of the type of tensile motion into his studies. But Galen went even further than this and developed a theory of equilibrium which makes a distinction between rest and the state of a body on which equal but opposing forces are applied. For this latter state he gives two illustrations: one is the swimmer against the current in a river who will remain in the same place "not because he does not move at all, but because he is carried forward

[43] Lucret., *De rerum nat.*, II, 308.
[44] Galen, *De musculorum motu*, I, 7-8 (Arnim, II, 450).

32

by his own movement by the same amount that he is carried back by the external movement".[44] The other is the picture of a bird suspended in mid-air because its upward thrust equals its weight. Galen is inclined to see equilibrium as a limiting case of the general one where one of the opposing forces is stronger than the other, thus resulting in a movement in the direction of the stronger force. The dynamical aspect of equilibrium leads him to an explanation in terms of tensional motion. He prefers to regard rest in this case as a purely phenomenological description of a rapid succession of movements in opposite directions which occur at such speed that to the observer the body seems to be in a state of rest.[45] Rest is in fact regarded here as rapid oscillation round the point of equilibrium. We see arising at this point, under the influence of Stoic concepts, a tendency to replace a merely vectorial point of view in mechanical considerations by a more elaborate dynamic notion.

3. *The Field of Force*

Foremost among the critics of tensional motion was Alexander Aphrodisiensis, whose arguments are of course motivated in the first place by his strict Aristotelian views. Tensional motion implies movement of the pneuma in opposite directions and therefore represents a serious violation of the concept of natural movement. It is unthinkable that pneuma should by itself move up as well as down in the course of its tensional motion within a body.[46] Alexander in raising this point in connection with his refutation of the Stoic theory of vision, adds another argument.[47] If the contact between the hegemonikon, as the centre of vision, and the outside world is really maintained by tensional motion of the pneuma, then vision could not be a continuous process but would be interrupted constantly during the time the impulse goes back to the hegemonikon in order to return again to the eye. The same should apply in fact—so he adds—to all the other kinds of sensation. In the age of cinematography Alexander would have refrained from making this objection, knowing that a rapid

---

[45] Aristotle was the first (cf. *Phys.*, 262 a 13) to state from a kinematic point of view that the reversing of a movement involves stopping it.

[46] Alex. Aphr., *De anima*, 131, 5 ff.

[47] loc. cit., 130, 26 ff.

succession of discontinuous events gives the impression of continuity. But even in his time Galen knew better, as we have just seen when he said that rapid oscillation round a point would not be distinguished from rest.

Another far-reaching Aristotelian objection is made by Alexander elsewhere,[48] when he accuses the Stoics of distorting Aristotle's conception of the aether. He claims that they misapply the notion of pneuma for similar purposes but contrary to the traditional peripatetic view. Superficially, it would seem that we have here a repetition of the former argument. Considering the ostensible affinity between pneuma and aether, an all-pervading pneuma contradicts the idea of the natural movement of the aether which is a circular one. But in fact the question at issue is a deeper one. In the Aristotelian cosmos, the unity and continuity of the universe were regulated and determined ultimately by the circular movements of the celestial spheres which were made up of the aetherial substance, whereas in Stoic physics this regulative power was given to the pneuma which was supposed to extend throughout the whole world and to create coherence by mixing with all matter. Here we have a significant and decisive transformation of the whole conception of the "fifth element" which was no longer restricted to a certain place but had become cosmic also in the topological sense of the word. As pneuma took over some of the characteristic attributes of aether, there began a confusion of these two terms and one was substituted for the other. Cicero says that "air resembles aether and is closely connected with it".[49] Arios Didymos expresses it directly: "Pneuma . . . has become analogous to aether, so that both are used synonymously."[50] Similarly, Clemens Alexandrinus referring to a passage in Aratos' poem, indicates that he would prefer to attribute the binding force of the cosmos to aether rather than to the Stoic tension of the pneuma (ch. I, n. 34).

These and other quotations suggest that it was the Stoic theory of pneuma which led to the change in the usage of the term "aether" which became of such historical significance. During the Middle Ages, when Aristotle reigned supreme, aether again became the element confined to the celestial spheres, but from

[48] Alex. Aphr., *De mixt.*, 223, 6 ff.
[49] Cicero, *De nat. deor.*, II, 66.
[50] Stob., *Eclog.*, I, 153, 24 (Arnim, II, 471).

the beginning of the seventeenth century on, the term was generally used to express the universal substance imbued with all the properties attributed to it by the Stoics, as is well known from the writings of Descartes, Boyle, Newton and others. There are of course some important differences between the pneuma of the Stoics and the aether of the physicists at the beginning of the modern era, differences which we will have to discuss. But the similarities are nevertheless striking, and it is this which gives special interest to the study of the notion of the pneuma and the criticism voiced against it. The significance of some of the basic scientific hypotheses and concepts is largely determined by their intrinsic functional and structural elements, irrespective of the historical period in which they took shape. It will therefore be worth while to analyse two more objections raised against the pneuma in antiquity. They will shed more light on the nature and the development of the concept of aether.

One of the objections is concerned with the apparent conflict between the tenuity of the pneuma and its cohesive function. From everyday experience—thus runs Galen's argument[51]—one should conclude "that the hard and rigid and compact is cohering by itself, whereas the tenuous and soft and yielding is in need of something else to make it cohere". Does it make sense to make fire and air, these very tenuous and soft substances, the cause of the hardness and rigidity of earth? How could they bestow on other substances those very properties which they themselves lack? Moreover, fire, far from binding things together, is known to have a dissolving power. Alexander argues along similar lines[52]: Pneuma is easily divisible and offers no resistance; "Some people have believed it to be akin to the void and of insipid nature, others to contain many empty spaces", and when mixed with other substances it would rather impart to them divisibility. On the other hand, if pneuma is really the cause of cohesion, then disintegrating and crumbling bodies should not contain any pneuma, and here arises the general question of the role of the pneuma with regard to substances which are coherent and can be divided nevertheless.

We have already seen that the genesis of the notion of pneuma supplies at least a partial reply to these questions. Its special

[51] Galen, *De multitud.*, 3, VII, 526 Kühn (Arnim, II, 440).
[52] Alex. Aphr., *De mixt.*, 223, 36 ff.

properties were derived from the combined qualities of air, the elastic substance, and of fire, the most active of elements. Air and fire also have a great pervading power, and in this connection we have to think not of the destructive properties of fire but of its activating characteristics, of the "innate heat" in organic bodies which at the time of Cicero was already regarded as one aspect of the moving power of heat in nature. It was the elasticity and the great pervasiveness of air, facilitated by its tenuity which, combined with the activity of heat, gave the pneuma all the qualities needed for a continuous medium and for a source of the cohesion of matter. But we cannot content ourselves with discussing the antecedents of the concept when analysing the point in question. We must remember that, although the Stoics believed in the corporeal nature of the pneuma, they came to regard it as something not akin to matter, but rather to force. It was their conception of a continuous field of force interpenetrating matter and spreading through space, and thus being the cause of physical phenomena, which formed the central idea of pneuma. This was a real innovation in the physical philosophy of the Greeks, as force was usually associated with impact and push, and as the old Empedoclean idea of the balance of power created by Strife and Love did not proceed beyond somewhat vague and poetically tinted descriptions. The idea of the existence of forces continuous in space and time merged in Stoic doctrine with the conception of the ever-present and all-permeating Deity. Pneuma became a concept synonymous with God, and either notion was defined by the other. On the one hand, natural force (i.e. pneuma) was seen as endowed with divine reason,[53] and pneuma was given epithets like "sensible" or "intellectual",[54] thus alluding to its god-like nature. On the other hand, God was identified with the all-penetrating pneuma,[55] being totally mixed with shapeless matter,[56] and divine reason was defined as corporeal pneuma.[57]

This way of looking on God and the active force of the pneuma as two aspects of the same agent clearly brings out the gist of the physical world of the Stoics. The cosmos is formed and ruled by

[53] Lactant., *Divina just.*, I, 5 (Arnim, II, 1025).
[54] νοερόν, ἔννουν; Plotin, *Ennead.*, IV, vii, 4; Aet., I, 6.
[55] Aet., I, 7, 33.
[56] Alex. Aphr., *De mixt.*, 224, 34 f.
[57] Origen., *Contra Celsum*, VI, 7 (Arnim, II, 1051).

forces which activate matter in a similar way to the activation of the living body by the soul. Sometimes the term *dynamis* is used to denote this force,[58] by virtue of which matter acts and moves.[59] If one adds to this the fact that in his theory of causes (cf. next chapter) Chrysippos defined Fate (*heimarmene*) as a pneuma-like force (*dynamis pneumatike*),[60] one comes to realize that pneuma derived its central position in Stoic physics from its dual significance. It was divine power (viz. Force) impressing a definite state upon matter on the one hand, and causal nexus linking the successive states of matter on the other, and in both these aspects it revealed itself as a spatially and temporally continuous agent.

It is evident that the identification of God and Force derived from the conception of an agent which had the faculty of shaping things and causing changes in the physical world. Force as a universal cause in this sense, according to the Stoic conception, shares with matter both extension and continuity, but it differs from matter in that it has very little substantiality (and bulk). The assumption of its extreme tenuity is in strict accordance with the hypothesis of its omnipresence not only within matter but also within the apparent emptiness of the space between the bodies. This aspect of pneuma is in a way the prototype of the field of force as it was developed in the physics of the nineteenth century, the significant difference being that in the age of mathematical physics this latter field concept was entirely stripped of any substantiality in the purely material sense of the word, nevertheless manifesting its physical reality by its effects on ponderable matter.

In many respects the field theory can be said to be an offspring of the aether hypothesis of the seventeenth century: the mathematical concept of the field and the forces prevailing in it took over some of the essential functions attributed to the aether. The aether of Descartes served as a medium to propagate motion across interstellar space; the aether of Boyle was supposed, in addition to that, to account for cohesion and "non-mechanical" phenomena like electricity and magnetism. Newton, finally, in the period of his life when he tried to supplement mathematical

---

[58] Cf. Alex. Aphr., loc. cit., 226, 12.
[59] Sext. Emp., *Adv. math.*, IX, 76.
[60] Stob., *Eclog.*, I, 79, 1 (Arnim, II, 913).

description by mechanical explanation, constructed a model of the aether to explain gravitation and optical phenomena. Some of his early ideas derive the aether from the principles of continuity and mixture in a picture surprisingly close to the basic Stoic assumptions: "Perhaps the whole frame of nature may be nothing but various contextures of some certain aetherial spirits and vapours, condensed as it were by precipitation . . . and after condensation wrought into various forms. . . . Thus perhaps may all things be originated from aether."[61]

Newton went even further and made conjectures about the aether as the origin of sensation. He speaks of "animal spirits of an aethereal nature subtle enough to pervade the animal juices . . .", and he assumed that "the coats of the brain, nerves, and muscles, may become a convenient vessel to hold so subtle a spirit". This statement is re-formulated very clearly in the famous last paragraph of the *Principia*: "And now we might add something concerning a certain most subtle spirit which pervades and lies hid in all gross bodies; by the force and action of which the particles of bodies mutually attract one another at near distances, and cohere if contiguous, and electric bodies operate to greater distances as well repelling as attracting the neighbouring corpuscles; and light is emitted, reflected, refracted, and heats bodies; and all sensation is excited, and the members of animal bodies move at the command of the will, namely by vibrations of this spirit, mutually propagated along the solid filaments of the nerves, from the outward organs of sense to the brain, and from the brain into muscles. . . ."[62] Here we see the hypothetical aether even taking over one of the essential functions attributed by the Stoics to the pneuma and its tensional motion within the body. However, one should not forget that every analogy has its limits, and one should bear in mind that Newton regarded matter as being of a porous structure. His aether, unlike the Stoic pneuma, therefore did not mix totally with matter, but acted within the pores, similar to Empedocles' effluences. Moreover, the functions of the pneuma were of a still more universal nature than that of Newton's aether, because, as we have seen (ch. I n. 39), it was supposed to be the cause of every conceivable

[61] Letter to Oldenburg (1675), quoted in D. Brewster, *Memoirs of the Life, Writings, and Discoveries of Sir Isaac Newton*, 1855, Vol. I, p. 392.
[62] Newton, *Principia*, ed. by F. Cajori, 1947, p. 547.

physical property of matter. A plausible explanation for this wider conception of the aethereal substance in Stoic physics can be found in the fact that the Stoic philosophy of nature was developed as a clear antithesis to the rival atomic philosophy of Epicuros. This question will be dealt with in a later section.

We shall now consider another objection against the pneuma which stems from peripatetic concepts, but whose implications are of a much more general nature. Alexander Aphrodisiensis, in his polemics against the Stoic theory,[63] argues that, if matter were being shaped through mixture with pneuma which in itself is a mixture of air and fire, there cannot exist in nature any simple body. He further questions the priority of the pneuma if one assumes it to be composed of two elements. By a similar train of thought, Simplicios[64] rejected the idea of the pneuma-like essence of the corporeal qualities, because of the composite nature of the pneuma; it would not, he says, produce coherence in other bodies if it possessed this property only by acquisition and not as something innate in itself. If we disregard the specific Aristotelian aspect of the argument, the point at issue is as follows: is it not basically inconsistent to assume the existence of a substance with such fundamental properties as the pneuma and at the same time to ascribe to it some sort of complexity? Is not structural simplicity necessarily the natural attribute of the aethereal medium? In answering this question we must distinguish carefully between the various functions of the pneuma. Pneuma as negation of the void, as physical expression of continuity, was no doubt conceived, if not as a simple substance, at least a homogeneous one. The assumption of its homogeneity, its sameness throughout the universe, made the question of its simplicity a secondary one as long as its main function was to pervade the emptiness of space. But pneuma as a force, as the cause of coherence and of all the physical qualities which differentiate inert matter, had by virtue of this function to be itself capable of some sort of differentiation. By necessity some functional co-ordination must be assumed between the various physical states and their causes. Every form of differentiation will, of course, put an end to homogeneity, but from that it does not follow yet that pneuma must be a complex substance. It could, for instance, be a substance capable of existing

[63] Alex. Aphr., *De mixt.*, 224, 15.
[64] Simpl., *Categ.*, 217, 36.

in different states of density, like the Air of Anaximenes, or like the aether in the later speculations of Newton, The latter example is especially illuminating, as Newton's idea was to explain the force of gravity by the gradually varying degrees of density of the aethereal medium. "Is not this medium much rarer within the dense bodies of the sun, stars, and planets and comets than in the empty celestial sphere between them? and in passing from them to great distances, doth it not grow denser and denser perpetually and thereby cause the gravity of those great bodies towards one another, and of their parts towards the bodies; every body endeavouring to go from the denser parts of the medium towards the rarer?"[65] When we try to find a closer analogy to the Stoic idea of the working of the pneuma, it will be worth while to contrast this principle of differentiation by density gradients with Newton's earlier speculations on the aether. In his letter to Oldenburg, quoted before, he says[61]: "But it is not to be supposed that this medium is one uniform matter, but composed partly of the main phlegmatic body of aether, partly of other various aethereal spirits, much after the manner that air is composed of the phlegmatic body of air intermixed with various vapours and exhalations. For the electric and magnetic effluvia, and the gravitating principle seem to argue such variety." The principle of differentiation by mixture propounded here is very similar to that of the Stoics which we could trace back to that of Diogenes of Apollonia (p. 10 f.). Pneuma, it is true, had gradually become the "fifth element", identified with aether,[50] and was probably in the course of time detached from its origins and regarded as a simple substance. But the Stoics, being in need of a differentiating mechanism, fell back upon its antecedents in order to take advantage of the two components fire and air. If we refer to the passage in Galen already quoted (ch. I, n. 9), we will realize that the principle of mixture was carried even further by assuming various degrees of dryness and humidity superimposed on the different compounds of hot and cold. To say that pneuma under certain circumstances is "wetter and colder" and sometimes "drier and hotter", comes very near to the Newtonian conception of a simple substance "intermixed with various vapours and exhalations".

[65] I. Newton, *Opticks*, Query 21. Cf. also Newton's letter to Boyle of Feb. 28, 1679, *The Works of R. Boyle*, 1772, Vol. I, p. cxvii.

4. *Cosmic Significance of the Pneuma*

The strong attacks of the later Aristotelians on the pneuma and its workings are significant, especially in so far as they reveal a reaction against the undermining effect which the Stoic theory undoubtedly had on certain Aristotelian concepts, and which made itself felt in the cosmological notions of followers and opponents of the Stoic School alike from the end of the second century B.C. on. These changes in the Aristotelian picture of the cosmos came as a result of the application of the theory of pneuma to the universe as a whole, although in the first instance this theory certainly enhanced and strengthened Aristotle's idea of the cosmic order. Indeed, the Stoic concept of *sympatheia*, the interaction and affinity of different parts of a unified structure (cf. p. 9) appears, when applied to the cosmos as such, as a logical extension of Aristotle's definition of Nature, as can be seen for instance in the first passage of Cleomedes' *De motu circulari*.[66] The first two attributes of Nature, he states, are the order of its parts and the order of its occurrences; the third is "the mutual interaction (sympathy) of its parts". Cosmic sympathy implies of course absolute non-existence of any vacuum within the cosmos, as Cleomedes asserts later on[67]: "In the cosmos there is no void as can be seen from the phenomena. For if the whole material world were not coalescent (*symphyes*) the cosmos would not be by nature coherent and ordered, neither could mutual interaction exist between its parts, nor could we, without one binding tension[68] and without the all-permeating pneuma, be able to see and hear. For sense-perception would be impeded by the intervening empty spaces." Similarly, Alexander says in the name of Chrysippos that Nature is made One by the pneuma which makes the Whole coherent and interacting (*sympathes*).[69]

The proof for the existence of sympathy by which the cosmos becomes a single body with a unified structure was seen in the influence that extra-terrestrial phenomena exert—or were held to exert—on the earth.[70] Among these the connection between

---

[66] Cleom., *De motu circulari*, I, 1 (ed. Ziegler), p. 2.
[67] loc. cit., I, 1 (p. 8).
[68] With Arnim I read τόνον instead of τόπον.
[69] Alex. Aphr., *De mixt.*, 216, 14. Cf. also loc. cit., 227, 8.
[70] Sext. Emp., IX, 79.

the moon and the tides is frequently mentioned, and of special significance are the passages in Cicero,[71] as they reflect the careful studies made by his teacher Poseidonios.[72] The periodic motions of the ocean are "in sympathy with the moon", and the tidal motions are thus regarded as a strong testimony for the prevailing forces of cosmic sympathy. Various explanations were given of the mechanism by which the attracting force is transferred from the moon to the seas,[73] but the essential point is that the Stoic theory for the first time implied a causal description of physical phenomena on a cosmic scale. Curiously enough, Poseidonios' idea of the constitution of the moon[74] was strongly influenced by orthodox Aristotelian notions of the difference between the sublunar regions and the pure aethereal realm above the moon. However, this did not prevent the conception of the universal significance of the pneuma and the notion of sympathy from revolutionizing the traditional view of the cosmic order. Slowly but firmly the conviction grew that the same laws prevailed everywhere in a cosmos permeated and ruled by one unifying pneuma. The first idea of a universal gravitation began to develop, suggested as it was in no small measure by the notion of the hexis and its application to celestial bodies. The conception that the moon and other extra-terrestrial bodies were held together by the cohesive forces of the hexis and have a structure of their own gave rise to a plurality of centres of attraction in preference to that of a single centre. This train of thought is very strikingly revealed in Plutarch's *On the Face in the Moon*, where the "terrestrial" properties of the moon [75] are emphasized by one of the disputants, an adherent of the Academy[76]: "If all heavy bodies converge to one and the same point, while each presses on its own centre with all its parts, it will not be so much *qua* centre of the universe as *qua* whole that the earth will appropriate weights, because they are parts of itself; and the tendency of bodies will be a testimony, not to the earth of its being the centre of the universe, but, to things which have been thrown away

---

[71] Cicero, *De divinatione*, II, 34; *De nat. deor.*, II, 19.

[72] Strabo, *Geographica*, III, 5, 8.

[73] Cf. e.g. Aet., III, 17, 5.

[74] Aet., II, 25, 5. Cf. also Plut., *D facie in orbe lunae*, 921 F.

[75] The Pythagoreans, and especially Philolaos, seem to have been the first to assign "earth-like" properties to the moon; cf. Aet., II, 30, 1.

[76] Plut., *De facie in orbe lunae*, 924 D.

from the earth and then come back to it, of their having a certain
and natural kinship (*symphyia*) with the earth. Thus the sun
attracts all the parts of which it is composed, and in the same way
the earth draws the stone to itself and makes it part of itself. . . .
But, if any body has not been allotted to earth from the beginning
and had not been rent from it, but somehow has a constitution
(*systasis*) and nature of its own, as they would maintain to be the
case with the moon, what is there to prevent its existing separately and remaining self-contained, compacted and fettered by
its own parts? For not only is the earth not proved to be the
centre, but the way in which things here press and come together
towards the earth suggests the manner in which it is probable
that things have fallen on the moon, where she is, and remain
there."

From the semantic point of view it is significant that the term
symphyia, which expresses a kind of dynamic continuity, is used
to denote the attraction by the earth, and that it is quoted in conjunction with *systasis* by Stobaios when he describes Chrysippos'
dynamics of the pneuma.[50] It is also interesting to note that the
expression used here for the pull or pressure of gravitation
appears in another of Plutarch's essays to describe the active
properties attributed to the Cold: frozen liquids exhibit the
cohesive forces of pressure contained in the Cold.[77] The destructive effect which the dynamic notion of pressure had on the
Aristotelian concept of one centre was partly restored by the
application of the hexis and its functions to the universe as a
whole. We have mentioned already (ch. I, n. 2) that the Stoics
regarded the cosmos as a finite body surrounded by an infinite
void. This enabled the cosmos to expand or contract according to
its various cosmic phases as described in the Stoic theory of
conflagration (*ekpyrosis*). The Aristotelian argument against the
Stoics was similar to the argument used against the Newtonian
cosmology, and claimed that a finite material universe would
scatter and dissipate into infinite space. Against this the Stoics
adduced the cohesive force of the hexis which the cosmos possesses
as a whole.[78] Through this hexis, all parts of the cosmos are
bound together into a "closed universe" which is not affected by
the void outside it, and thus the character of a single coherent

[77] Plut., *De primo frigido*, 946 B.
[78] Cleom., *De motu circulari*, I, 1, (p. 10).

entity is given to the universe, again bearing a closer resemblance to the Aristotelian cosmic hierarchy.

5. Continuum Theory and Atomic Theory

The Stoic doctrine of pneuma was the first consistent and elaborate continuum theory of matter. It is a counterpart to the first atomic theory of matter which was conceived about one hundred and fifty years earlier (Leucippos taught about 450 B.C. and Zeno about 300 B.C.) and reached its climax a few decades before the development of the Stoic School (Epicuros lived about sixty years before Chrysippos). Like the atomic theory, the continuum theory of the Greeks was essentially speculative, based on theoretical conceptions and developed along purely epistemological lines. Although both theories occasionally refer to experience and use examples and analogies borrowed from the sphere of daily life, there is no question of any recourse to systematic experimentation.

It might be worth while to compare briefly the principles on which the two rival systems are founded. The comparison of these principles is indeed of interest for the study of the formation of physical concepts within the framework of theories that represent a self-consistent system of thought, even though lacking a proper experimental foundation. It appears that there is an intrinsic logic in these systems leading to analogies and associations some of whose principal features we find again in analogous theories of the age of experimental science and mathematical physics.

Each of these two theories was based on two fundamental entities—the atomic theory on the atoms and on the void,[79] and the continuum theory on the unformed matter (*hyle*) and on the pneuma. The mutual relation of these entities in each theory immediately determines the reciprocal character of the one with respect to the other. The theory of atoms postulated the total separation of atoms and the void: "The atoms are solid and have no share in the void."[80] The atoms were thought to be filled out completely with substance, a fundamental supposition accentuated by the term "complete *plenum*".[81] The possibility of a

---

[79] Galen, *De elem. sec. Hippocr.*, I, 2 (Diels, 68 B 125): ἄτομα καὶ κενόν.
[80] Simpl., *De caelo*, 242, 18.
[81] Arist., *De gener. et corr.*, 325 a 29: παμπλῆρες.

mixture of "void" and "solid" would imply a porous structure of the smallest particles, whatever they may be, and this would lead in the last instance to a negation of the basic assumption of the theory, to a "crumbling away" and dissolution of matter, as was explained by Epicuros.[82] The continuum theory, in contradiction to this, postulated the total mixture of hyle and pneuma. The term "without share" quoted above from Simplicios[80] is used by Alexander with the opposite meaning, when he discusses the interpenetration of soul and body: "There is no part of the soul which has no share in the body which contains the soul."[83] Dissolution of the body sets in after the soul has left it, and the same applies to the pneuma in general: a slackening of its intensity of penetration leads to a relaxation of the binding forces of the hexis.

Let us now consider the principles which, in each theory, lead to a differentiation of the bodies with regard to their physical qualities. In the theory of atoms there are the three well-known elements of shape (*schema*), order (*taxis*) and position (*thesis*),[84] typical of the discrete character of the theory. Shape specifies how "solid" is outlined against the "void" and represents a basic geometrical distinguishing mark between the various kinds of atoms. We know from Theophrastos that in Democritos' theory most of the secondary qualities were essentially determined by the shape of the atoms.[85] Order and position, too, are geometrical characteristics, but they elaborate the relationship between solid and void by introducing the principle of grouping as a further differentiating mechanism. By the various arrangements in space of equal or different shapes the possible transformations of quantity into quality are thus brought to completion. In the continuum theory the counterpart of shape is proportion. In a given mixture of pneuma and matter the actual ratio of the components of the pneuma—fire and air—determines a definite differentiation (ch. I, n. 9) and corresponds to a definite shape. As against the group principle of order and position in the theory of discrete particles, we have in the continuum theory the super-position principle resulting in the hexis. We have seen that hexis,

---

[82] Epicur., *Letter to Herodotus*, 56.
[83] Alex. Aphr., *De mixt.*, 217, 36.
[84] Arist., *Metaph.*, 985, b 15.
[85] Theophr., *De sensu*, 65-67, 73-75.

the physical structure of a body, is nothing else but the super-position of all the mixtures of pneuma corresponding to the various qualities of the body. Each of these mixtures forming the ensemble of the hexis co-exists with the others, and they all together permeate the body as tensional motions, thus making it a dynamic entity. The assumption of continuous forces of tension within the body represents the most significant difference between the two theories. There is no continuous interaction between the atoms of Democritos; they are not surrounded by fields of force as in the modern theory, because the Greek atomists denied the existence of such forces. They only admitted interaction by direct contact,[86] either through impact or by mechanical interlocking.[87]

It cannot be denied that in the later period of the atomic theory there were certain indications of a development of concepts similar to that of a vibration of atoms within a larger unit ("molecule"). Some passages in Lucretius admit of such an interpretation[88] and make it plausible that an important step was taken with the assumption of co-ordinated interaction within an ensemble of discrete particles. But it should be noted that this interaction was a purely kinematical one and excluded any continuous forces. It is worth while mentioning that Aristotle had already stressed the difficulties of a theory based on principles of arrangement and order alone in the case where one looks at the same body in different states of aggregation.[89] How is it possible to explain the difference between water and ice by *thesis* and *taxis* alone? To what an extent the Stoic notion of forces and tensions prevailing within matter has influenced scientific views in this field, one can learn from Plutarch's essay on the Cold[90] which has already been mentioned above. Cold is by no means inert, and the apparent immobility of a cold body like ice is actually the expression of the solidity of its structure, having its origin in "a force which displays a cohesive and binding tension".

We can complete the comparison of the two theories by mentioning their different ways of describing the propagation of phenomena in space. Up to the time of the Stoics, not only the

[86] Arist., *De gener. et corr.*, 325 a 33.
[87] Simpl., *De caelo*, 295, 11.
[88] Lucret., *De rerum nat.*, II, 111; II, 1009.
[89] Arist., *De gener. et corr.*, 327 a 15.
[90] Plut., *De primo frigido*, 946 b-d.

atomists but all natural scientists knew of the locomotion of
bodies only: Aristotle talks of "motion in respect of place".[91]
The term "translatory motion"[92] is put against "rotatory
motion"[93] in a passage mentioning Heracleides' hypothesis of the
axial rotation of the earth.[94] Philo, on the other hand, has already
been quoted by us[31] as putting "translatory motion" against
"tensional motion" whereby the former term is expressly characterized as locomotion. There can be no doubt that the conception of the propagation of a state was developed for the first time
by the Stoics, and it is not by chance that they also were the first
to describe the expansion of sound in space in the form of
spherical waves.[5]

In conclusion, we must point out that Stoic continuum physics
and the atomic theory of the Greeks differed in their attitude
towards a principle which has played an important role in the
philosophy of the modern era—the principle of the identity of
indiscernibles. The atomists emphasized the complete equality of
all atoms of the same kind: "The nature of them all is one, just
as if each one separately were a piece of gold."[95] The plausible
inference drawn from this assumption was that two different
macroscopic bodies, too, would be completely equal provided they
were composed of atoms equal in kind and arrangement. This
view was also shared by opponents of the atomic doctrine such
as the Peripatetics, who maintained the possibility of the absolute
indiscernibility[96] of two different bodies.[97]

On the other hand, the Stoics insisted that "nothing is the same
as that which some other thing is" (nihil esse idem quod sit aliud),
and that "no hair or grain of sand is in all respects the same as
another hair or grain".[98] Although the argument on this point
with Carneades and his followers centered mainly round the
subjective aspect of the problem (i.e. the human ability or inability to distinguish between false and true sensations and to
tell apart two apparently equal bodies), the Stoic opinion had

---

[91] Arist., *Phys.*, 265 b 25.
[92] μεταβατικὴ κίνησις.
[93] τρεπτικὴ κίνησις.
[94] Aet., III, 13, 3.
[95] Arist., *Ce caelo*, 276 a 1.
[96] ἀπαραλλαξία.
[97] Cicero, *Acad. pr.*, II, 50 f. and 54; Plut., *De comm. not.*, 1077 c.
[98] Cicero, loc. cit., II, 85; cf. also Plut. loc. cit.

actually deeper roots. It stemmed from their strict determinism which was but another aspect of their continuum conception and which will be dealt with at length in the following chapter. The Stoics held that if two bodies can be individually recognized as being in two different places, the causal nexus which has thus placed them differently is a sufficient reason for their being different in some respect. In fact, their complete indistinguish-ability would mean that they are identical.

It was no mere chance that Leibniz, who re-stated the *principium identitatis indiscernibilium*, also firmly adhered to the continuum doctrine (his monads differ from each other and vary through a continuous range). In the fourth letter of his correspond-ence with Clarke he writes: "There is no such thing as two individuals indiscernible from each other. . . . Two drops of water or milk, viewed with a microscope, will appear distinguishable from each other. This is an argument against atoms."[99] And in his fifth letter he reiterates: "I infer from that principle . . . that there are not in nature two real, absolute beings, indiscernible from each other. . . . This supposition of two indiscernibles, such as two pieces of matter perfectly alike, seems indeed to be possible in abstract terms; but it is not consistent with the order of things, nor with the divine wisdom, by which nothing is admitted without reason."[100]

Modern quantum physics, however, has shown that electrons and other elementary particles satisfy the principle of identity of indiscernibles by obeying the Fermi-Dirac Statistics based on Pauli's exclusion principle. This latter principle is in fact an extension of the idea of the Stoics and of Leibniz into the world of discrete particles and basically statistical happenings.

[99] *The Leibniz-Clarke Correspondence*, ed. H. G. Alexander (New York, 1956), p. 36.
[100] loc. cit., p. 61.

III

THE SEQUENCE OF PHYSICAL EVENTS

1. *Cause and Effect*

FROM the concept of the continuum and the dynamics of its parts which the Stoics developed, they managed to advance and to make significant progress in the epistemology of the causal nexus. The history of modern physics has taught us that the analysis of the problem of causality has been decisively assisted and furthered by two basic methods which were foreign to Greek antiquity—systematic experimentation and the mathematization of science. The refinement of experimental techniques on the one hand revealed the complexity of apparently simple phenomena and on the other hand made it possible to decompose them into a number of isolated occurrences. It could further be shown that some of these occurrences were connected with others, and that a clear and unambiguous description could be given of the connections between them. The language most suited for this description is mathematics, and the success of mathematical physics became indeed an overwhelming proof of the possibility of describing causally increasingly large complexes of the physical world. Mathematical algorithms, such as differential equations, tensor analysis, matrix algebra and statistical methods, have been successfully used or adapted to describe the ever-growing sum total of our knowledge, to assist in the detection of new facts and to serve as instruments of prediction.

This very fact has firmly established in our consciousness the identity of natural law and causality and has given the idea of causal nexus a well-defined and scientific meaning. Although the Greeks could hardly have had even a bare inkling of this specific notion of causality, there are definite trends of development in their formation of causal notions, and a large share in this is due to the Stoics.

The central concepts of natural law in the pre-Stoic period were *ananke* (necessity) and *aitia* or *aition* (cause). The former appears, for instance, in the well-known fragment from Leucippos "nothing happens at random; everything happens out of reason and by necessity".[1] *Ananke* here stands for natural law, and indeed the whole Greek philosophy of nature since Thales is a paraphrase of the cognizance that cosmic happenings are subject to regularities and therefore exclude any arbitrariness. The very fact that ordered experience is possible was taken as a proof for the existence of Cause. Statements to this effect probably date back to early times. We find them mentioned by Sextus Empiricus in the name of the Sceptical School: "If cause were non-existent everything would have been produced by everything at random."[2] Absence of cause would result in lack of permanence in all phenomena—a horse could be formed from a man, a plant from a horse, things proper to summer could happen in winter, and vice versa, etc.

A similar argument was used by Aristotle in order to prove that Nature belongs to the class of causes which he called "final".[3] It must be emphasized, however, that Aristotle's classification of causes into material, formal, efficient and final ones[4] was not conducive to the methodology and epistemology of ancient physics. The final cause, which expresses a purpose for change, acquired an overwhelming preponderance over the other types of causes. The result was that the teleological approach which held no key to the physical world largely determined the physical sciences, and especially dynamics, till the seventeenth century. In Aristotle's "efficient cause", which he defines as "primary source of change", we can see some hint of the heuristic function of the causal law as an instrument for the discovery of causal

---

[1] Diels, 67 B 2.
[2] Sext. Emp., *Pyrrh. hyp.*, III, 18; cf. also *Adv. math.* IX, 202 ff.
[3] Arist., *Phys.*, 198 b 10 ff.
[4] loc. cit., 194 b 15 ff., *Metaph.*, 983 a 24 ff., 1013 a 24.

connections in Nature. Some aspects of the "formal cause", on the other hand (e.g. "the ratio 2:1 and number in general are causes of the octave"[5]), contain the nucleus of the idea of the mathematical formulation of natural laws.

While Aristotle's formalism was of little influence on the further elucidation of the notion of causality, practical medicine, already long before his time, paved the way for future development through the application of diagnostics and aetiology. Ample proof for this is supplied by the treatises of the Hippocratic Corpus. Clinical observation must consist of "declaring the past, diagnosing the present, foretelling the future"[6] —here we have in a nutshell, dating back to the fifth century B.C., the expression of a causal attitude and a prescription based on determinism. The word used for "cause" in the Hippocratic writings is usually *aitia*, but occasionally *ananke* appears in a synonymous sense.[7] *Ancient Medicine* treats at length of the investigation of possible causes which may lead to various diseases. This remarkable essay shows clearly how medical practice leads straight to the core of the problem of causality. The physician is confronted in his aetiology with the difficulties arising from the multiplicity of causes and from the time lag between the dominant cause and the effect. On the one hand there is the dependence of symptoms and their intensity on various kinds of food as well as on the constitution of the patient[8]; on the other hand, the author points out the tendency of physicians and laymen alike to assign the cause of a disturbance to an event antedating the effect by the smallest time lag, which can very easily lead to a wrong diagnosis.[9] We find a clear refutation of the non-causal or "spontaneous" (*automaton*) event in another essay belonging to the Hippocratic Corpus, called *The Art*. The passage is quoted here because the statement on the validity of the causal law is backed by reference to medical experience and to the sources of medical prognosis: "No patient who recovers without a physician can logically attribute his recovery to spontaneity. Indeed, under a close examination spontaneity disappears. For everything that occurs will be found to do so through something, and this

---

[5] *Phys.*, 194 b 27, and *Metaph.*, 1013 a 28.
[6] Hippocr., *Epidemics*, I, xi, 9.
[7] Hippocr., *Airs, Waters, Places*, XXI.
[8] *Ancient Medicine*, VI, XI, XX.
[9] loc. cit., XXI.

'through something' shows that spontaneity is a mere name and has no reality. Medicine, however, because it acts 'through something' and because its results may be forecasted, has reality."[10]

The theoretical approach to the doctrine of cause of the Hippocratic School of medicine which looked for the causal factors of organism and environment in order to explain and cure the disease resulting from them was later somewhat neglected by Erasistratos and the "Empiricists". Against them Galen also, 600 years after Hippocrates, stresses again the necessity for the practitioner (*iatros*) to act as a scientist (*physikos*): one has to investigate the causes which upset the functions of the organs in order to be able to bring them back to normal.[11]

The tradition established by the medical discipline of repeated observation of symptoms in patients and of relating them to factors prevailing before the outbreak of the disease and during the disease helped to bring about the Stoic theory of causality, which constitutes their contribution to the problem. The contribution of the Stoics to this chapter in the epistemology of science can be summed up in the following four points: (1) They made a more penetrating analysis of the cause-effect relation which brings it nearer to the idea of natural law, and they extended it to a general deterministic pattern; (2) within the frame of their interpretation of mantics, they were the first to state the connection between causal law and induction; (3) within the framework of causality they investigated the notion of the Possible and its relation to the Necessary; (4) they accomplished the first transition from causal thought to functional thought.

Once again we must deplore the loss of the earlier Stoic writings which forces us to reconstruct their ideas from later sources— Stoic and other— mainly from Clemens Alexandrinus, Alexander Aphrodisiensis, Sextus Empiricus, Simplicios, Plutarch, Cicero and Seneca. From these sources there emerges the picture of a theory of causality which is largely determined by the dynamic notion of continuity dominating the physics of the Stoics, and by their conception that everything capable of acting and being acted upon is a body.[12] Space and time are necessary conditions

[10] Hippocr., *De arte*, 6.
[11] Galen, *De nat. facult.*, II, ix.
[12] cf. e.g. Sext. Emp., *Adv. math.*, VIII, 263.

of every causal connection.[13] As physical events are transmitted by nearby action, either through direct contact of bodies or by the pneuma, this must be true also for cause-effect relations. Contiguity is therefore an essential attribute of causality, and causes are bodies acting upon other bodies either in spatial contact with them[14] or through the medium of the pneuma.[15] The nature of this action can always be described as movement, as there are always motions involved in it, either the tensional motion of the pneuma or locomotion of other bodies. These motions are therefore common to the bodies acting and acted upon,[16] and it is they that are to be regarded, according to the Stoics, as effects resulting from essentially bodily causes: "Every cause is a body which is the cause to a body of something incorporeal happening to it."[17] The examples which follow this quotation throw into relief the functional character of the cause-effect relation: the effect can be expressed as a verb, such as "being cut" in the case of lancet and flesh, "being burnt" in the case of fire and wood.

Here we have a radically new approach to causality as against Aristotle and all the precursors of the Stoics. Instead of the vague formulation "A is the cause of B" (in the language of today), the Stoic definition elaborates thus: A is the cause of the effect E being wrought on B. According to this statement the Stoics saw effect as a process originating in a body A and leading to a change in a body B. The course taken by the process, the direction from A to B, is an expression of the second attribute of causality, in addition to contiguity, namely antecedence. Both contiguity and antecedence are in fact nothing but specifications of the *conditio sine qua non* of the spatio-temporal structure of causality mentioned above.

A result of extreme significance follows for the Stoics from their concretization of causes as bodies, in conjunction with their doctrine of the complete community in which all bodies exist by

---

[13] Clemens Alex., *Stromat.*, VIII, 9 (Arnim, II, 346): τὰ ὧν οὐκ ἄνευ; Seneca, *Epist.*, 65, 11.

[14] Simpl., *Categ.*, 302, 31: πελάσει καὶ ἄψει. Cf. also Procl., *In Parmenid.* V, 74 (Arnim, II, 343).

[15] Aet., I, 11, 5.

[16] Simpl., *Categ.* 306, 14: τὸ κοινὸν τοῦ ποιεῖν καὶ πάσχειν εἶναι τὴν κίνησιν.

[17] Sext. Emp., *Adv. math.*, IX, 211: πᾶν αἴτιον σῶμά φασι σώματι ἀσωμάτου τινὸς αἴτιον γίνεσθαι.

virtue of the dynamic medium, the pneuma. They realized that in every given instance one has to reckon with a multiplicity of causes, since the complex texture of natural phenomena reduces the conception of one single body acting upon another to a mere abstraction. Taking account of this complexity they arrived at a formulation of the law of causality which was transmitted to us by Alexander as follows: "In view of the multiplicity of causes, they (the Stoics) equally postulate about all of them that, whenever the same circumstances prevail with regard to the cause and the things affected by the cause, it is impossible that sometimes the result should be this and sometimes that; otherwise there would exist some uncaused motion."[18] This postulate comes remarkably near to our present notion of causality. Today we are aware of the fact that the causal law in its strict sense can be applied only to systems which can be (nearly) isolated from the rest of the world and which can be subjected to recurrence, i.e. where the same constellation can be (nearly) restored. With these qualifications we could postulate that if a state A leads to a state B, a state A' closely similar to A will lead to a state B' closely similar to B. The deficiency of the Stoic definition derives from the fact that the conception of artificial isolation and wilful recurrence—both of which have sprung from the technique of systematic experimentation—were unknown to Greek science. However, it should be noted that the Stoic statement, which simply reads "every time A is restored B must follow again", is the first statement on causality on record which introduces the element of recurrence and the idea of reproducibility of a situation B from a situation A. This implies the possibility of the prediction of events and thus leads straight from causality to determinism, to the concept of *heimarmene* (fate).

It is interesting to see that some of the objections raised in modern times against the usual formulation of the causal principle were voiced already in antiquity in the criticism directed against the Stoic view. A conspicuous example can be found in Alexander Aphrodisiensis' book *De fato*.[19] Alexander rejects the unqualified connection of the notion of causality with the regular succession of an event B upon A. "We see that not all the events succeeding each other in time occur because of those which

---

[18] Alex. Aphr., *De fato*, 192, 21.
[19] Alex. Aphr., *De fato*, 194, 25 (ch. 25).

occurred before them . . . neither is night caused by the day . . . nor summer by the winter." The criteria for the non-causal character of the relation of day and night have been discussed by Mill and others.[20] The relevant fact that the light of the day is always connected with the sun is brought as an argument by Alexander: "We observe that the mutual order of day and night has one and the same cause, as has also the change of the seasons. For winter is not the cause of summer, but the common cause of both is the revolution of the divine body and the inclination of the ecliptic in which the sun moves, this being the cause alike of all the phenomena mentioned." In justice to the Stoics we may assume that they would have argued as Alexander did; in the case of the succession winter-summer, at least, evidence for this can be found in their definition of the seasonal changes as due to the movements of the sun.[21]

Before turning to the Stoic doctrine of determinism, we must focus our attention on the last sentence of their law of causality which refers to the "uncaused motion"[18] and which is ostensibly directed against the Epicurean idea of the spontaneous, uncaused swerve of the atoms. According to the Stoic conception, every transition from the state A, characterized by a certain constellation of bodies, to another state B, happens by way of motions that represent the effects of the causes leading to that transition. As all these motions are on the one hand linked by the causal nexus to A and on the other hand do inevitably lead to B, a situation C different from B could only arise if there were at least one motion different from the others. But such a motion would not be accounted for by the sum of all causes represented by the state A and would therefore be acausal. Such an uncaused motion is inconceivable in the Stoic world governed by the strict continuity of the pneuma and its dynamics, where nothing can be added to or subtracted from the sum total of occurrences. As a result of this trend of thought, the Stoics were led to identify strict causality with some kind of principle of conservation: "They say that an uncaused event resembles a creation *ex nihilo* and is just as impossible."[22] We have here before us, in a qualitative

[20] Cf. H. Weyl, *Philosophy of Mathematics and Natural Science* (Princeton, 1949), p. 193.
[21] Diog. Laert., VII, 151.
[22] Alex. Aphr., loc. cit., 192, 14.

way, an anticipation of the connection which Newtonian mechanics established between the fundamental laws of dynamics (which are a mathematical expression of the law of causality) and the laws of conservation of momentum and energy. Moreover, the passage quoted alludes also to the doctrine of the "conservation of the existing", formulated by the Atomists,[23] which excludes every creation from nothing.

The Stoic argumentation against the causeless event was not restricted to purely theoretical deliberations; it also referred to experience, as the following quotation shows: "Chrysippos confuted those who would impose lack of causality upon nature by repeatedly referring to the astragalos and the balance and many other things which cannot alter their falling motions or inclinations without some cause and variance occurring in themselves or outside them."[24] The game of dice is possibly mentioned here because of its common usage in antiquity and because it was evident to everyone that dice would only fall if thrown. However, by mentioning the balance Chrysippos might have been referring to the laws of the lever discovered by Archimedes who was Chrysippos' contemporary. Postulates 2 and 3 of his *Equilibrium of Planes* start from the equilibrium of the balance and regard its inclinations as effects of the increase or decrease of one of the weights. This statement leaves no doubt that there exists a cause of the perturbation of the equilibrium. In all those cases where such a cause cannot be discovered by any external symptom, Chrysippos, excluding any chance,[25] postulates its existence thus: "For there is no such thing as lack of cause, or spontaneity. In the so called accidental impulses which some have invented, there are causes hidden from our sight which determine the impulse in a definite direction."[24] This is in complete agreement with the definition of chance as "hidden cause", often attributed to the Stoics.[26]

We have seen that the Stoics regarded cause as a body and effect as a movement affecting another (or the same) body. As this body in turn can be the cause of other effects, one arrives at

---

[23] Cf. e.g. Democritos in Diog. Laert., IX, 44: μηδέν τε ἐκ τοῦ μὴ ὄντος γίνεσθαι μηδὲ εἰς τὸ μὴ ὂν φθείρεσθαι.

[24] Plut., *De Stoic. repugn.*, 1045 c.

[25] Or "spontaneity"; cf. Arist,. *Phys.*, II, ch. 4-8.

[26] τύχην αἰτίαν ἄδηλον ἀνθρωπίνῳ λογισμῷ; cf. Aet., I, 29, 7 (Arnim, II, 966); Simp., *Phys.*, 333, 3; Alex. Aphr., *De fato*, 174, 2.

the conception of a chain of causes stretching continuously in space and time and thus forming in their totality the course of the universe. "From everything that happens something else follows depending on it by necessity as cause, and every event has a forerunner, the cause upon which it depends."[27] The reason for this is easily found in the continuum doctrine: " . . . because there is nothing in (the cosmos) which is separated and divorced from all that happened before." Here we see another aspect of the uncompromising attitude of the Stoics towards the Epicurean *clinamen* (deviation) of the atoms. The elimination of one single link from the causal chain would by necessity lead to the destruction of the chain as a continuous whole. The concatenation of causes[28] establishes an interdependence whose disconnection would amount to a dissolution of the cosmos. "The cosmos would break up and be scattered and could not longer remain a unity administered by one order and plan, if some uncaused movement were to be introduced into it."[27]

2. *Determinism and Free Will*

The close analogy between the dynamic continuum representing the Stoic cosmos and their notion of the causal nexus is clear. Effects are propagated in space and time and, together with the bodies from which they emanate and which they affect, form the totality of causation. Thus by virtue of the topology and dynamics of the causal nexus, the doctrine of causality expands into determinism, the doctrine of *heimarmene*. In the pre-Stoic period, *heimarmene* was mainly used to denote human fate,[29] but the Stoics introduced it as a term signifying causality, the ordered system of causal occurrences. The Stoic etymology of *heimarmene* (from *eiro*, string beads), indicating this notion, is used in one of their numerous definitions of Fate: "*Heimarmene* is the continuous (literally: strung together) cause of things."[30] Similarly, another source quotes the following definition: "*Heimarmene* is the ordered interdependence of causes."[31] Still another definition, quoted as being by Chrysippos, comprises both order and inter-

---

[27] Alex. Aphr., *De fato*, 192, 6 ff.
[28] Plotin, *Ennead.*, III, i, 2: ἐπιπλοκὴ πρὸς ἀλλήλων τῶν αἰτιῶν.
[29] Cf. Pauly-Wissowa, "Heimarmene."
[30] Diog. Laert., VIII, 149.
[31] Aet., I. 27, 3: συμπλοκὴν αἰτιῶν τεταγμένην.

dependence: "Heimarmene is a natural order of the Whole by which from eternity one thing follows another and derives from it in an unalterable interdependence."[32]

From the concept of order and interdependence in the topological sense embodied in the definition of *heimarmene* follows its meaning as *logos*, as the divine order and law, by which the cosmos is administered,[29] and "by which the past events have happened, the present ones are happening and the future ones are going to happen".[32] On the other hand, the dynamic nature of *heimarmene*, the succession of causes and effects by contiguous action, is expressed either by the identification of fate with a pneuma-like force (*dynamis pneumatike*),[33] or directly through *kinesis*: "an eternal motion, continuous and ordered".[34] Cicero sums up several of these definitions in a passage[35] to which we will refer in greater detail when we come to discuss the Stoic theory of induction. It is, however, illuminating to compare this passage in Cicero with the famous Laplacean definition of determinism, as such a comparison proves the striking conformity of their conception with his. The sentence in Cicero reads thus: "By fatum I mean that which the Greeks call *heimarmene*, that is, an orderly succession of causes wherein cause is linked to cause and each cause itself produces an effect. . . . Therefore nothing has happened which was not bound to happen, and, likewise, nothing is going to happen which will not find in nature every efficient cause of its happening. . . . If there were a man whose soul could discern the links that join each cause with every other cause then surely he would never be mistaken in any prediction he might make. For he who knows the causes of future events necessarily knows what every future event will be."[35] And here follows Laplace's description of the omniscient intellect: "We must therefore regard the present state of the universe as an effect of the state preceding it, and as cause of the state which will follow it. An intellect which at a given moment would know all the forces governing Nature and the respective situations of all things of which it is composed . . . would embrace in the same formula the movements of the largest bodies in the universe as

---

[32] Gellius, *Noct. Att.*, VII, ii, 3.
[33] Stob., *Eclog.*, I, 79, 1 (Arnim, II, 913).
[34] Theodoret., VI, 14 (Arnim, II, 913).
[35] Cicero, *De divinatione*, I, 125-128.

well as those of the lightest atom; nothing would be uncertain for it, and future and past alike would be present before his eyes."[36]

There is also a significant parallel in the reference to astronomy: "Persons familiar with the rising, setting and revolutions of the sun, moon, and other celestial bodies can tell long in advance where any one of the bodies will be at a given time."[37] And here are Laplace's words: "The human mind, by the perfection to which it has succeeded in bringing astronomy, represents only a faint idea of this supreme intellect."

Laplace was able to base his opinion on the differential equations and the perturbation theory of astronomy the results of which were in such striking agreement with experience. His determinism was founded on a sweeping and bold extrapolation of the achievements of celestial mechanics which had succeeded in describing causally, i.e. by mathematical equations, a certain section of physical reality. The solution of these equations represented in general terms the orbits of bodies that obey the natural law regarded as the "cause" of the observed phenomena. This general solution could be applied to a particular case when the initial conditions of that case were known and introduced into the solution. Initial conditions simply mean the position and velocity of each of the bodies concerned at any chosen moment, and they define the "state" of those bodies at "zero hour".

Only observation can furnish the initial conditions, for instance the position and velocity of a planet at a given moment, but after these data have been introduced into the theoretical solution, the final result allows the determination of the state of the bodies at any other time in the past or the future.

In antiquity there existed nothing similar to this marvellous system of celestial mechanics developed in the seventeenth and eighteenth centuries which could have corroborated the Stoic belief in absolute determinism. Neither were the mathematical tools available to formulate even the simplest laws of kinetics nor was Greek mentality inclined towards that analytical attitude to nature which is the prerequisite for the development of systematic experimentation. It was all the more remarkable consequently that the Stoics tried to build a conceptional frame suited to serve

[36] P. S. Laplace, *Essai philosophique sur les probabilités*, I (Gauthier-Villars, 1921), p. 3.
[37] Cicero, loc. cit., 128.

their theory of determinism and capable of expressing their ideas about the causal description of events. Here medical notions, already developed from the time of Hippocrates (cf. p. 51), served as a pattern and a starting-point for further development— another instance of the frequent trend of Greek thought to begin from biological concepts in the basic description of Nature, even of the inorganic world.

Greek medicine had always emphasized the necessity to look for the "immediate starting-point" of an illness, and again Hippocrates can be quoted in this respect.[38] By this starting-point the whole history of the disease is divided into two parts—the chain of events leading up to its outbreak, and its course proper which is largely determined by the disposition of the patient. Accordingly physicians were led finally to distinguish between two types of cause, "antecedent causes" (causae antecedentes) and "active and operating causes" (causae activae et operantes). This distinction is known to have been described in the writings of two physicians, Erasistratos, a contemporary of Chrysippos, and Asclepiades of Prusa who lived at the time of Cicero.[39]

These two causes form part of the Stoic classification of causes which replaced that of Aristotle. Among the sources concerned with this, Clemens of Alexandria is the most informative.[40] In addition to antecedent and operative causes the Stoics introduced two further types: the "intensifying (synergon) cause" was defined as one which without being able by itself to produce an effect, intensifies the effect of the operative cause in a similar way to that in which synergists are defined in modern chemistry. Finally, the term "joint cause" (synaition) was used to characterize the additive nature of certain causes which together bring about the resulting effects. The division of causes into four types seems to have been made in analogy to that of Aristotle. However, only the first two of them are essential for the understanding of the Stoic approach to causality and fate.

The most important difference between the antecedent and the operative cause according to Clemens Alexandrinus is that the effect of the first persists after the cause has ceased to exist,

[38] Hippocr., *Epidemics*, II, 1, 12: "One has to look for the starting-points (ἀφορμάς) of his illness."
[39] M. Pohlenz, *Die Stoa*, vol. I, p. 105; vol. II, p. 61.
[40] Clemens Alex., *Stromat.*, VIII, 9.

whereas the operative cause and its effect co-exist only in strict simultaneity, so that the effect sets in with the cause and ceases together with it.

This distinction found a very interesting application in the passionate discussion of free will versus fate which was caused by the extreme deterministic position of the Stoic School. On the one hand it was clear that free will and fate are mutually exclusive notions, as was bluntly stated by the Peripatetic Diogenianos[41]: "I believe that in the same way as the notions sweet and bitter, black and white, hot and cold are absolute contraries, so are also 'by our free will' and 'by fate'; provided one defines 'according to fate' as the course things would take under any circumstances, whether we follow readily or not, and 'by our free will' a course which will reach its goal through our endeavours and actions, but will fail to reach it if we do not care and are idle." On the other hand, as the negation of free will would obviously destroy the foundations of ethics, the Stoics were forced to look for some reconciliation within their deterministic system. Chrysippos' attempts in this direction, however unsatisfactory they may appear to the metaphysical mind, are of special significance to the historian of physical thought because of their epistemological implications.[42]

The whole causal chain leading up to a certain situation and ending in a definite act on the part of a man confronted by that situation is divided by Chrysippos into two parts. The first part comprises all the external events and includes as a last link their presentation to the conscious human mind. These links in their totality form the antecedent causes[43] whose deterministic nature is obvious and indisputable. The attitude taken by the mind to the situation presented to it, the *appetitus* which is either an irrational impulse or a reasoned choice and which results in the subsequent act forms the second part of the chain which constitutes the operating causes, sometimes also called "perfect and principal causes".[44] Now the gist of Chrysippos' theory is that the

[41] Euseb., *Praep. evang.*, VI, 8, 30.

[42] The main sources for what follows are Cicero, *De fato*, 39 ff.; Plut., *De Stoic. repugn.*, 1056 b ff.; Alex. Aphr., *De fato*, ch. 13; Nemes., *De nat. hom.*, ch. 35.

[43] According to Cicero (loc. cit., 41) they were also called "auxiliary and proximate" (adiuvantes et proximae).

[44] Cicreo, loc. cit.: perfectae et principales, and Plut., loc. cit.: αὐτοτελεῖς αἴτιαι (complete in themselves, independent).

perfect causes are "within our power",[45] and thus establish freedom of will. This assertion is illustrated by a simile which goes back to Chrysippos and is found in the literature in two variants. Cicero[46] quotes Chrysippos as pointing at the cylinder and spinning-top which cannot begin to move unless they are given an impulse; but once the impulse has been given, the cylinder goes on rolling and the top spinning according to their own nature (suapte natura). A similar illustration is quoted by Aulus Gellius[47]: "If you roll a cylindrical stone over a sloping steep piece of ground, you do indeed furnish the beginning and cause of its rapid descent, yet soon it speeds onward, not because you make it do so, but because of its nature and because of the ability of its form to roll (quoniam ita sese modus eius et formae volubilitas habet)." The quotation continues by explaining the simile: "Just so the order, the law and the necessity of fate set in motion the classes and beginnings of causes, but the impulse of our design and thought and the actions themselves are determined by each individual's own will and the characteristics of his mind."

The logical consistency of Chrysippos' formal train of thought within the deterministic framework can be elucidated by comparing it with the formalism of Newtonian analytical dynamics today: let us consider the simple act of the motion of a body; it is causally determined by a differential equation into which enter the co-ordinates, their (first and second) derivatives with respect to time and certain material constants, and by the initial conditions, i.e. the data fixing the position and velocity of the body at a certain moment. The solution of the equation gives the orbit of the body in general terms, without reference to a definite physical situation. By introducing the initial data (furnished by the knowledge of the boyd's state at zero hour) into the general solution, the state of the body is determined for all other times before and after this given moment. We may say, using the Stoic terminology, that the initial conditions represent the antecedent cause, whereas the equation itself and especially the physical constants entering it represent the principal cause. Both are part and parcel of the deterministic scheme and the body's

---

[45] ἐφ' ἡμῖν, literally "with us".
[46] Cicero, loc. cit., 42.
[47] Gellius, *Noct. Att.*, VII, ii, 11.

fate depends on both of them, but there is a difference between them. The initial conditions are the last link of the whole history of the world which led to this particular state of the body. The stone in Chrysippos' second simile was doubtless subject to various tribulations before "zero hour", when it started rolling down the slope. Being put into a certain situation is the result of an extremely complicated chain of causal happenings comprising the whole history of the physical world preceding that situation. It is clear that the part played by the stone itself in this antecedent chapter is insignificant compared with all the other events which finally created the initial conditions. But if we come to regard the particular behaviour of the stone during its descent down the slope, we realize that the course of its motion is determined by the principal or perfect causes among which the physical pro- perties of the body play an important part. In the case of the stone, it is, as Chrysippos expresses it, its "modus" and "formae volubilitas"; in the case of a vibrating string it is its modulus of elasticity and its density; in the case of the spinning top—its moment of inertia; in a moving fluid—its viscosity, and so on. All these parameters, although they are themselves inseparably bound up with the whole "fate" of world events, are nevertheless characteristic quantities of the body itself, and it is through them that the body's own share in its determined course becomes manifest.

In Chrysippos' simile the physical parameters of the material body correspond to the "main characteristics" of the human mind. For him, the human impulse to act is an innate cause of motion in man, exactly as gravity is the cause of the natural motion of the stone—both perform their actions "by necessity but not against their nature".[48] Alexander also quotes these words and expounds this aspect of Chrysippos' theory in great detail in the thirteenth chapter of his book *On Fate*. From his account and from Nemesios' discussion of the subject it is evident that Chrysippos' argument was based on the Aristotelian concept of natural movement. The various natural tendencies of things— inanimate and living alike—are determined by fate. As long as no external impediment opposes the natural downward motion of the stone, the upward motion of the fire, the fruitbearing of plants or the impulse of living beings to act according to their

[48] κατ' ἀνάγκην οὐ τὴν ἐκ βίας : Alex. Aphr., *De fato*, 181, 23.

own will, these natural movements will be performed in accordance with the strictly determined parameters, i.e. according to the perfect or principal causes. Any interference from the outside, from the antecedent or proximate causes, will lead to a resultant motion which is no longer a "natural" one, but which of course is still determined by fate. This solution of the antithesis between free will and fate was still defended by the Stoic Philopator, nearly four hundred years after Chrysippos,[49] but outside the Stoic School it was generally rejected as a fallacy. The Cynic Oinomaos, a contemporary of Hadrian, called the lot accorded to man by Chrysippos "semi-slavery",[50] and both Alexander and Nemesios remark that in the last instance the theory amounts to complete determinism, free will being reduced to a mere play with words. Nemesios in his argument with Chrysippos uses the Stoic definition of universal causality already quoted by Alexander[18]: "If in the case where the same circumstances prevail everything must happen exactly in the same way, as they say, and not sometimes one way and sometimes another because it was so allotted since eternity, then the impulse of the living being, too, must necessarily and absolutely arise in the same way if the same causes prevail."[51] Real free will, he continues, means that in the same circumstances it is in our power to yield to the impulse or not to yield. This last remark is elucidated further by a passage in Plutarch[52] which tells us about a situation regarded by the opponents of the Stoics as an experimental proof of the reality of "true" free will, i.e. the existence of an adventitious impulse[53] in man: "If one has to choose between two things of equal worth and exactly alike, and there is no cause which drives one towards either of them, because they differ in nothing, then the very adventitious impulse of the soul itself decides and breaks the deadlock."

Chrysippos refused to accept such an overthrow of the psychological equilibrium as proof of an accidental (i.e. capricious) impulse and replied with the argument already quoted earlier[24] which insists on the strict deterministic character of human volition and attributes apparent fortuity to "hidden causes".

[49] Cf. Nemes., loc. cit.
[50] Euseb., *Praep. evang.*, VI, 7, 2 and 14.
[51] Nemes., loc. cit.
[52] Plut., *De Stoic. repugn.*, 1045 b.
[53] ἐπελευστικὴ κίνησις.

In Alexander's and Nemesios' criticism of the physical approach of the Stoics to the problem of free will we find pronounced, quite particularly, their resentment of the analogy drawn between the "physical constants" of inanimate and of living bodies. The identification of the human impulse with natural movement, says Nemesios, raises the obvious question which is in fact a *reductio ad absurdum*: Does fire burn of its own free will because it is its nature to burn? He then adds a semantic analysis indicated by Alexander: the Stoics seem to confound two concepts—"in our power" (literally: "with us"), and "through us".[54] One cannot see in the performance of a certain act through an agency an expression of the latter's free will: "By the same reasoning it would be the free will of the lyre or the flute and the other instruments and all the animals and inanimate things, if somebody operates through them—which is absurd."[55]

In fairness to the Stoics one can, against this criticism, sum up once more their essential point as follows: Both the "personal parameters" of man and the "initial conditions" are part of an all-embracing fate. But by the very fact that the personal parameters participate in the determined chain of events *independent of external conditions,* the arbitrariness of fate is alleviated. Psychologically this participation appears in the consciousness of the human mind as free will.

3. *Divination and Induction*

Although the strict validity of the causal nexus within the physical world was regarded by the Stoics as axiomatic truth, they nevertheless endeavoured to give it the support of empirical evidence, as we have seen in the instance of Chrysippos' argument against the "causeless movement".[24] Since Galileo and Newton, the (classical) concept of causality has been able to establish itself firmly on the result of both systematic experimentation, especially the breaking up of complex phenomena into isolated and reproducible events, and of the mathematization of physics. In ancient Greece, the Stoics had to look for essentially observational sources of evidence in order to provide proofs of their theory of universal

[54] ἐφ' ἡμῖν and δι' ἡμῖν. Sometimes παρ' ἡμᾶς is used as a synonym for ἐφ' ἡμῖν.
[55] Nemes., loc. cit.

causality. As experiments connecting phenomena in the realm of the physical sciences were scarce, they were led to incorporate into their evidence the vast field of divination and to make use of all the available facts and interpretations in favour of their doctrine. The introduction of this highly controversial subject into physics gave rise to a prolonged and sometimes very heated debate with the opponents of divination, especially with the Peripatetics, a debate which reveals the great importance attributed by the Stoics to divination as an illustration of the principle of induction and as a proof of the law of causality.

The Stoic belief in "artificial" divination, i.e. in divination from inference based on signs and events in the physical surroundings, in animal life, etc., stemmed from their attitude towards "natural" divination, i.e. the alleged faculty of certain inspired persons to foresee the future. In this they followed in the footsteps of the Pythagoreans and Plato, who had defined divination as "the mutual association of Gods and men".[56] This association or fellowship which the Pythagoreans extended even to animals was due according to them to the world soul which pervades the whole universe,[57] and in the Stoic doctrine this belief was turned into a physical explanation in their theory of the pneuma: "Since the universe is wholly filled with the Eternal Intelligence and the Divine Mind, it must be that human souls are influenced by their contact with divine souls."[58] It should be noted, however, that already in pre-Stoic times the gift of prophecy was sometimes regarded not merely as a transcendental faculty, restricted to a certain class of selected people, but also as an art based on rational foundation, and therefore accessible to every intelligent being.[59] Antiphon even defines divination as "the conjecture (*eikasmos*) of a wise man",[60] whereby *eikasmos* obviously hints at an inference by comparison with similar cases.

This second kind of divination, the artificial or inductive one, became of great importance in Stoic epistemology, as can be seen in great detail in Cicero's two books *On Divination* and to some extent in the fragments of the Peripatetic Diogenianos (probably

[56] Plato, *Symp.*, 188 b.
[57] Sext. Emp., *Adv. math.*, IX, 127.
[58] Cicero, *De divinatione*, I, 110.
[59] Plato, *Phaedr.*, 244 c.
[60] Diels, *Fragmente der Vorsokratiker*, 87 A 9; cf. also Cicero, loc. cit., I, 116.

between the first century B.C. and the first century A.D.) which are preserved for us by Eusebios.[61] Cicero clearly defines inductive divination as "the art of those who follow up new things by inference, having learned the old ones by observation".[62] The empirical basis of divination which was established through long-continued and repeated observation of signs, is again and again stressed by the Stoic School.[63] All events are predetermined, and the course of time is only unfolding them in their order like the unwinding of a coil of threads. As certain combinations and constellations of events repeat themselves continually, careful study of their nexus can furnish means of knowing the future.[64] Basing themselves on their strictly deterministic doctrine, the Stoics denied any essential difference in method between scientific inference and inductive divination: meteorology, for instance, also attempts to forecast storms by certain signs of the weather.[65] The obvious objection to this attitude, namely that science (for instance medicine) is rational, whereas the origins of divination are not intelligible,[66] was not recognized as valid. Magnetism is a well-known phenomenon, but would it not be foolish to deny magnetic attraction because there is no explanation for it?[67] Far from taking an apologetic stand on this point, the Stoics, in a spirit similar to that of modern science, firmly insisted on the positivistic approach to phenomena. The main task of a scientist is to correlate results and to stick to "observables" rather than to inquire into the nature of causes.[68] We are by no means certain about the causes of those meteorological phenomena from which we infer the coming of storms, but we will understand their effect and therefore rightly rely on them.[69] The same applies to divination: even though we may not discern the causal nexus itself, we still discern the signs and tokens of those causes.[70]

These arguments and similar ones give some indication as to the Stoic attitude which came to regard divination as a legitimate

---

[61] Euseb., *Praep. evang.*, IV, 3, 1-13.
[62] Cicero, loc. cit., I, 34.
[63] e.g. Cicero, loc. cit., I, 2, 12, 25, 109, 127.
[64] loc. cit., 127-128.
[65] Cicero, loc. cit., I, 13.
[66] Cicero, *De nat. deor.*, III, 15.
[67] Cicero, *De divinatione*, I, 86.
[68] Cicero, loc. cit., I, 12.
[69] loc. cit., I, 16.
[70] loc. cit., I, 127.

"science, based on observation and interpretation",[71] or, as another source puts it, as "an art (*techne*, which thus can be acquired by skill), based on the evidence of certain results".[72] From the same source we learn that Zeno and Chrysippos as well as Poseidonios were strict adherers of this opinion, with only Panaitios dissenting and denying divination any real basis of truth.

The quotations given above are perhaps not sufficient in themselves to give a plausible explanation for the firm stand of the Stoics on the subject of science versus divination in the face of the strong attacks and striking counter-arguments brought up by such an eminent philosopher as Carneades (and put on record in Cicero's book *On Divination*). The issue was, however, much more fundamental for the Stoics, since it concerned the problem of interdependence of causality and induction. "According to the Stoics, God is not present at every little fissure in the liver or in every song of a bird . . . but the universe in the beginning was so instituted that certain events are preceded by certain signs. . ."[73] In other words, the law of causality supposes an immanent and pre-established order in the world by which the succession of single events and the interconnection of phenomena, including those related through divination, is determined once and for all. It follows, therefore, that the postulation of determinism (i.e. fate) implies the assumption of the validity of the method of induction (i.e. divination).

Diogenianos quotes the same statement from Chrysippos' book *On Fate*: "The predictions of the diviners could not be true if Fate were not all-embracing."[74] On the other hand, the fact that divination "worked", and that the diviners had their successes, was a strong argument in favour of causality. Each prediction that came true added a new instance to the sum total of all the cases noted by experience where the event *B* followed every time that the factual situation *A*, as seen by the diviner, repeated itself. Thus for the Stoics the ever-lengthening chain of fulfilled predictions increased the belief in future successes and represented an experimental proof for determinism based on the

[71] Sext. Emp., *Adv. math.* IX, 132: ἐπιστήμη θεωρητικὴ καὶ ἐξηγητική.
[72] Diog. Laert., VII, 149.
[73] Cicero, *De divinatione*, I, 118.
[74] Euseb., loc. cit., IV, 3, 1.

generalization of the observed concatenation of the events A and B. Diogenianos remarks further: "Chrysippos has given us a proof based on the mutual dependence of things. For he wants to show by the truth of divination that everything happens in accordance with fate; but he cannot prove the truth of divination without first assuming that everything happens in accordance with fate." There seems to us no justification for Diogenianos' rejection of the Stoic method of "reciprocal inference", which he obviously looked on as a vicious circle. On the contrary, Stoic epistemology is here in full accord with the approach of modern science, which is firmly based on the mutual confirmation of the causal law and the principle of induction, as it manifests itself for instance in the interplay of experiment and mathematical deduction in physics ever since Newton.

There remains, of course, the basic question to what extent the predictions of the diviners come true. Even the adherents of divination conceded that not all predictions are fulfilled.[75] On the other hand, the most passionate objectors did not deny that some events were correctly predicted.[74] The sometimes heated argument on this point, which is reflected in Cicero's De divinatione and in the fragments of Diogenianos, sheds further light on the firm stand of the Stoics in the matter of divination, and at the same time contains the first discussion on whether a result has significance. The Stoic position is that of the empirical scientist who has to allow for errors of observation. "Signs badly guessed and badly interpreted turn out to be false not because something is wrong with the order of things but because of the ignorance of the interpreters."[76] Failures of this kind occur in every empirical science containing an element of conjecture, as for instance in medicine.[77] Again we notice the contention of the Stoics that there is no difference in principle between divination and science and as both deal with the forecasting of determined events, every discrepancy which becomes apparent *post factum* has to be attributed to the inadequacies of the experts or the imperfection of their methods.

However, Diogenianos in his polemic against divination completely rejects the idea that those discrepancies are the result of

[75] Cicero, loc. cit., I, 124.
[76] loc. cit., I, 118.
[77] loc. cit., I, 24 (Cf. also Cicero, De nat. deor., II, 12).

"errors of observation". According to him, only a small fraction of the predictions come true, proving that divination is not art at all and that the occasional hits are mere coincidences. Diogenianos in his analysis[61] adopts the Aristotelian theory of chance and incidental cause[78] to the problem in question, and especially Aristotle's distinction of "rule" and chance: "Rule applies to what is always true or true for the most part, whereas chance belongs to a third type of event."[79] Starting from this conception Diogenianos arrives at a hard and fast rule for the significance of a result: Chance (*tyche*) does not mean a complete lack of success, but it means not succeeding in all cases or even in most cases. The occasional successes of the diviners have therefore no scientific basis but they are the work of a chancelike cause (*tychike aitia*).[80] This contrasting of science and chance is found already in an earlier passage where Diogenianos sarcastically asks if one would call somebody who hits the mark only once a skilled archer, or one who kills most of his patients a physician. Here *episteme* is used in the sense of "skill" in a similar way to that by which Aristotle defines experience (*empeiria*) as something akin to science (*episteme*) and art (*techne*),[81] referring thereby to Plato's saying that experience produces art and inexperience chance.[82] But the full significance of Diogenianos' argument can only be understood if his use of *episteme* is seen against the background of the passage in Aristotle's *Physics* quoted above and the parallel passage in *Metaphysics*.[83] Divination is not a science nor an art where perfection can be acquired by skill, because it pertains essentially to that category of events which have an accidental character, i.e. which are not determined in principle and whose realization cannot be predicted but only registered *post factum*. Tyche in the sense of the lucky coincidence of prediction and fulfilment is only a special case of an *aitia kata symbebekos*, of an accidental cause,[83] and there is no science of the accidental, as Aristotle points out.[84] According to Diogenianos and the Peripatetics, the failures of the diviners in the majority

[78] Arist., *Phys.*, II, 5.
[79] loc. cit., 197 a 18. Cf. Also *Metaph.*, 1026, b 32 ff.
[80] Euseb., loc. cit., IV, 3, 6: οὐκ ἐπιστήμης ἀλλὰ τυχικῆς αἰτίας ἔργον.
[81] Arist., *Metaph.*, 981 a 2.
[82] Plato, *Gorg.*, 448 c.
[83] Arist., *Metaph.*, 1064 b 33—1065 a 6, 1065 a 30-35.
[84] loc. cit., 1027 a 20.

of their predictions only confirm the basic essence of the world as a contingent, not strictly deterministic one. Only in such a world —if at all—would divination make sense as a guidance in practical decisions; in the strictly deterministic world of the Stoics knowledge of the inexorable future could only add to the suffering of the individual.[85] Diogenianos rejects divination in principle with all the arguments of a rational scientist, which in his time, or a few decades before him, were summed up by Carneades.[86] One should, however, remember that the consultation of diviners was not only common practice in antiquity, but for long periods even an official institution and a public custom. The Stoics probably had an answer to the question how to reconcile the custom of divination with their belief in fate. The words quoted by Cicero: "all that is going to happen to us is made more bearable when the religious rites are fulfilled",[87] are perhaps only one of the answers. But in conclusion it must be again emphasized that the whole problem of divination occupied the Stoics not so much for practical reasons but was of primarily theoretical or scientific interest to them, for the very reason that they had to accept the validity of divination in a deterministic world and at the same time saw in it a confirmation of determinism by inductive inference.

4. *The Role of the Possible in the Deterministic Scheme*

The revolution created in Greek scientific thought by the deterministic theory of the Stoics can best be illustrated by the change it brought about in the notion of the Possible and the associated discussion. Aristotle's point of view which became dominant in later antiquity and in the Middle Ages is evident from several passages in his *De Interpretatione*[88] and *Metaphysics*[89] and from the analysis of his commentators, especially the very lucid exposition in Boethius' commentary to *De Interpretatione*.[90] Aristotle does not always draw a clear line between the logically possible and the empirically possible, and he uses at

---

[85] Euseb., loc. cit., IV 3, 7 f.
[86] Cicero, *De divinatione*, II 9 ff.
[87] loc. cit., II, 25.
[88] Arist, *De interpretatione*, chs. 9 and 13.
[89] Arist, *Metaph.*, 1019 b 21 ff., 1046 a 7, 1047 a 21-1047 b 30, 1050 b 11-15.
[90] Boeth., *Comment. in De Interpret.*, 2nd part, ed. Meiser.

the same time, for illustrating purposes, sentences such as "it is impossible that the diagonal (of a square) is commensurable" and "it is possible that a man should be seated". His discussion of propositions concerning future events reveals Aristotle as an out-spoken non-determinist, believing in the existence of a real contingency in the world. The possible was for him not only the opposite of the impossible but also of the necessary, in the sense that whereas necessary events are bound to happen with certainty nothing can be predicted about a possible event which might or might not occur at some future date. This division of events into necessary and contingent ones is intimately connected with Aristotle's hierarchy of physical phenomena within the cosmos, as Boethius clearly explains[91] in his comment on Aristotle's well-known analysis of the sentence "tomorrow there will be a naval battle".[92]

Predictions concerning the celestial bodies, such as "the sun will set today, the sun will not set today", can be divided immedi-ately into true and false ones, because the heavenly phenomena are, as they are, by necessity, and this immediately excludes one of the opposites. However, this is not the case with propositions about things that are going to happen in the sublunar sphere and that are of a transient nature ("quae in generatione et corrup-tione sunt"), such as "Socrates will read today" or its opposite "Socrates will not read today". It is obvious that one of the alternatives must be true, but which is true and which false remains in abeyance until the time stated ("today" in the example quoted) has elapsed. The same reasoning applies of course to the sentence about the naval battle. Here also the attributes "true" and "false" have no meaning until after the time limit of tomorrow. This, according to Boethius' interpreta-tion of Aristotle's views, has nothing to do with our capacity to recognize future events but springs from the nature of the event itself.[93] Events happening in the sublunar sphere, and particu-larly everything pertaining to the field of human action, are therefore contingent, i.e. undetermined in principle, in contra-distinction to the eternal recurrences of the heavenly motions

---

[91] Boeth., loc. cit., p. 244 ff.

[92] Arist., *De interpret.*, 19 a 28 ff.

[93] Boeth., loc. cit., p. 197: "Peripatetici non in nobis hoc sed in ipsa natura posuerunt."

and the successions of seasons depending on them which are "of necessity" and can therefore be predicted.[94] In connection with this, further light is shed on Aristotle's rejection of determinism in his statement (made in another context) on chance (*tyche*)[95]: "Causes from which chance results might happen are indeterminate; hence chance is obscure to human calculation and is a cause by accident." Expressed in the language of modern determinism this would mean that according to Aristotle even a Laplacean intelligence could not predict events of the sub-lunar world because they are contingent by their very nature, and may happen thus or otherwise.

Out of the intimate connection of the notion of the possible with that of the accidental, i.e. the non-necessary, there emerged another important point on which Aristotle was very definite. If some event *A* has the nature of a contingency, it necessarily must happen at least once, because if it will never happen it would have the characteristic of the impossible, and its opposite, non-*A*, would be a necessity: "It cannot be true to say that a thing is possible but will not be."[96] In other words: if we always called events possible which never came true, then anything might be possible, and the term "impossible" would lose its sense. The same point is taken up from another aspect in chapter VIII of the *Physics*, in the discussion of motion.[97] The motion of the spheres is an eternal necessity, because if it ceased and motion were replaced temporarily by rest, this motion would be accidental.

The same train of thought we find in the definition of the Possible given by the Megarian Diodoros Cronos, a contemporary of Aristotle. It was a definition most widely accepted in the post-Aristotelian era and down to the late Hellenistic period, and it is quoted by several authors such as Cicero, Plutarch, Alexander and Boethius.[98] This definition is given as "The possible is that which either is true or will be true" (Cicero, Plutarch), or, in a slightly less complete form: "The possible is that which either is or will be" (Alexander, Boethius). Here we have in a concise formulation all the essential characteristics of the notion attributed to the

[94] Arist., *De gener. et corr.*, II, ch. 11.
[95] Arist., *Metaph.*, 1065 a 33 f.
[96] Arist., *Metaph.*, 1047 b 4-5.
[97] Arist, *Phys.*, 256 b 10.
[98] Cicero, *De fato*, 13; Plut., *De Stoic. repugn.*, 1055 e; Alex. Aphr., *Anal. Pr.*, 183, 34; Boeth., *De interpret.*, p. 234.

possible by Aristotle. Either it is already realized, i.e. in a state of actuality, or else, within the sphere of contingent events, it has to be realized at some future date, because if it were to remain infinitely in a state of potentiality, it would be impossible and its opposite alternative will come to bear the character of the necessary.

The disciple of Diodoros, the Megarian Philo, however, took another view in this matter, as is reported by Simplicios[99] and mentioned also by Boethius.[98] "How shall we decide (says Simplicios) whether something is perceptible (*aistheton*) or knowable (*episteton*)? Shall we decide by the fitness alone, as Philo said, even if there is no knowledge of it, nor ever will be? As for instance that a piece of wood in the Atlantic Ocean is combustible in itself and according to its own nature?" We have here a definite shift of the concept of the possible into another logical category, whereby possibility is seen as an application of the definition of a certain property to a special case. We know by experience that wood can be burnt and, having arrived by induction at combustibility as a property of wood in general, we ascribe it as a possibility also in cases which may to the best of our knowledge never be realized. A less radical version of this definition is given in the continuation of the passage quoted: "Or shall it rather be decided by the unhindered (*akolytos*) fitness according to which a thing can be subject to knowledge and perception by itself as long as no manifest obstacle stands in the way?" Here "fitness" is used not in the "bare" sense[100] but in the qualified one of "unhindered", meaning that the external circumstances should be such that they do not prevent the possible from being realized. Boethius is obviously quoting this version when he says: "Philo says that something is possible if by its internal nature it is susceptible of truth, as for instance if I say that today I shall read again Theocritos' *Bucolica*. This, if nothing external prevents it, can be truly stated, as far as it concerns its inner nature." The essential difference between Philo's criterion of the possible in this latter version and that of Diodoros or Aristotle is thus that Philo does not stipulate the actualization of the possible event at any future date, provided that this actualization is not prevented by external circumstances.

A deep gulf separates the Stoic approach to the possible from

[99] Simpl., *Categ.*, 195, 31 ff.   [100] ψιλός; cf. Simpl., loc. cit., 196, 11.

that of their predecessors, in spite of the formal similarity of their definition with that of Philo, who, by the way, was a friend of Zeno the Stoic. This need not surprise us in view of the deterministic conception of the Stoics which *prima facie* leaves no room for the possible and restricts any future alternative to the opposites "necessary" and "impossible". The Stoic definition which, according to Diogenes Laertios, goes back to Zeno is as follows: "A proposition is possible which admits of being true, there being nothing in external circumstances to prevent it being true."[101] A similar version is to be found in Boethius,[98] and a rather elliptical one is given by Cicero,[98] which, however, makes up for this shortcoming by the illustration added to it: "You (Chrysippos) say that things which will not be are also possible— for instance it is possible for this jewel to be broken even if it never will be." The following passage from Plutarch is probably characteristic of the opinion generally held in the Hellenistic period as to the impossibility of reconciling this statement— which indeed sounds very much like Philo's—with determinism: "How could there be no contradiction between the doctrine of the possible and the doctrine of fate? If indeed the possible is not that which either is true or will be true, as Diodoros postulates, but everything is possible that admits of coming true though it may never come about, then there will be many things possible among those which will not happen in accordance with unconquerable, unassailable and victorious Fate. Either the power of Fate will dwindle or, if Fate is as Chrysippos supposes it to be, that which admits of happening will often become impossible. For all that is true will necessarily be, being compelled by supreme necessity, but all that is false will be impossible, the strongest cause preventing it from becoming true." [102]

Obviously the Stoics had to discard the former notion of the possible which signified an objective contingency in a non-deterministic world, and to replace it by something compatible with determinism. They did so very logically by making it a subjective category and basing it on human ignorance of the future. "According to them (the Stoics) the possible is relative to our cognition," reports Alexander, [103] and he adds that for those

---

[101] Diog. Laert., VII, 75.
[102] Plut., *De Stoic. repugn.*, 1055 d-e.
[103] Alex. Aphr., *De fato*, 176, 26.

who are able to know the full causal nexus, as for instance the diviners, the possible does not exist. The meaning of "chance" underwent a change which was linked up with the new significance of the possible. "The Stoics who believe that everything happens out of necessity and by providence judge the casual event not according to the nature of chance itself but according to our ignorance."[104] It is interesting to note that the Aristotelian definition of *tyche* quoted earlier[95] was taken over literally by the Stoics[105] and thereby the "obscurity to human calculation" was given a new meaning. Again we learn from Alexander[106] that "the assertion that chance is a cause obscure to the human mind is not a statement about the nature of chance but means that chance is a specific relation of men towards cause, and thus the same event appears to one as chance and to another not, depending on whether one knows the cause or does not know it". Alexander, amplifying on the Stoic view, adds that by this logic things occurring within the frame of a science or an art will appear as the result of chance to the untrained and unprofessional.

The Stoic conception of the possible as arising from human ignorance of the future in a deterministic world, on the one hand led to a deeper comprehension of causality, and on the other clarified for the first time the problem of disjunctive propositions containing statements subject to empirical verification. The account given by Alexander[107] is particularly illuminating in this respect. He quotes the Stoic definition of the possible as follows: "The possible event is something that is not prevented by anything from happening even if it does not happen." The significance of this version becomes evident when it is considered in connection with the theory of disjunctions which was developed extensively in Stoic logic.[108] Here we are interested in the exclusive disjunctions, i.e. the disjunctions which are true if just one of their component members is true ("either A or B or C . . . is true"), and among these we are concerned with those of an empirical content which can be verified or proved false only through immediate experience or at some future date. It is obvious that, within the deterministic framework, only one of

[104] Boeth., loc. cit., 194, 23.
[105] Simpl., *Phys.*, 333, 3.
[106] Alex. Aphr., *De anima*, 179, 6 ff.
[107] Alex. Aphr., *De fato*, ch. 10 (176, 14 ff.).
[108] Mates, *Stoic Logic* (University of California Press, 1953), p. 61 f.

the alternative, *A, B, C*, . . . will be realized, whereas the rest will not come true. But the essential point is that *all of them are possible* if only "they are not prevented by anything from happening", i.e. for instance if, as far as our knowledge goes, none of them are in contradiction to the rules of nature. Here the Stoics take a stand similar to that of outspoken determinists in the modern era. The cases which are not prevented from happening are analogous to the "equally possible cases" of Laplace, who defined them as cases about whose existence we are in equal ignorance or as cases about which we have no reason to assume that one of them would be realized rather than the others.[109]

The Stoic doctrine of the possible amounts therefore to a widening, or generalization, of the deterministic scheme. Instead of seeing causation as a one-dimensional chain of actual occurrences they saw it as a many-dimensional network of potential occurrences, all of them equal possibilities fitting within the frame of "fate", out of which, and in accordance with the rules of disjunction, only one course will be actualized. It seemed perfectly justified to the Stoics on logical grounds to include in their deterministic network a case *A* which later might not be realized: "Those things which are the causes for the opposite things to happen according to fate are also the causes for the non-happening of the things themselves." In other words: the same causal nexus which led to the happening of *B* (which was supposed to be the exclusive disjunction of *A*) was the reason which prevented *A* from happening, and therefore both *A* and *B* have their place in the causal scheme.

Having thus fitted both the possible and its opposite, the impossible into the deterministic schema, the Stoics, with their keen sense for clear-cut terminology, endeavoured to put the term "necessary" in its proper place. "Tomorrow there will be a naval battle" is a *possible* proposition (as is its opposite), but it is not a necessary proposition, even should it prove to be true.[110] The attribute "necessary" has to be reserved for propositions which are *always* true, such as logical or mathematical statements. "This (statement about the naval battle), however, does not remain true after the battle has taken place." Time-dependent

---

[109] P. S. Laplace, *Théorie analytique des probabilitées*, 1812, p. 178-179.
[110] Alex. Aphr., loc. cit., 177, 7 f.

propositions, therefore, which either lose their significance after the realization of the event or prove to be false in the case of non-realization, are possible propositions. If the naval battle takes place, it will happen as a possible event which at the same time is also a determined one. "It is not inconsistent that an event should be a possible one within the sum total of events which occur according to fate."

From another source[111] we know that this view of the necessary was not held by Chrysippos who disagreed on this point with his teacher Cleanthes. Chrysippos argued that "all propositions true about the past are necessary, because they are unchangeable and because things past cannot turn from true into false". This statement is part of an argument which was conducted round a famous logical lemma, the so called "Master" of Diodoros[112] which must be briefly mentioned here because it has indirectly a bearing on Stoic epistemology. Diodoros argued that only two of the following propositions could be true:

(1) All propositions true about the past are necessary.
(2) An impossible proposition does not follow from a possible one.
(3) A proposition which neither is true nor will be true is yet possible.

We recognize in the third proposition the Stoic definition of the possible; it was accepted, according to Epictetos, by Chrysippos as well as by Cleanthes who disagreed about the first two, Chrysippos accepting the first and Cleanthes the second. Diodoros, on the other hand, chose the first two and used them to prove his conclusion about the nature of the possible.[98]

Chrysippos' choice of the first and third proposition led him into difficulties in view of his positive attitude towards divination, as Cicero tells us in a rather facetious way.[113] The instance chosen is a conditional proposition based on the observation of astrologers: "If anyone is born at the rising of the dogstar, he will not die at sea." Let this statement be applied to a special case: "Fabius was born at the rising of the dogstar, therefore Fabius will not die at sea." The first part, belonging to the past, is a necessary proposition according to Chrysippos; the second part

[111] Cicero, *De fato*, 14.
[112] Epictet, *Diss.*, II, 19; cf. Mates loc. cit., p. 38 f.
[113] Cicero, loc. cit., 12-16.

(the true prediction) in consequence becomes also necessary and its opposite (the false statement "Fabius will die at sea") an impossibility. Thus Chrysippos would be forced to accept Diodoros' definition of the possible by identifying it with either what is or what will be true. According to Cicero, Chrysippos extricated himself from this difficulty by giving the conditional proposition the form of a negated conjunction: "Not both: someone is born at the rising dogstar and he will die at sea." Leaving aside the subtleties of the purely logical argument, the essential point to be considered here is the obvious endeavour of Chrysippos to avoid conditional formulation in empirical propositions related to future events. No doubt he would have preferred to restrict the rather apodictic conditional to molecular propositions of *logical* content, i.e.—to use the Stoic expression—to propositions which are *always* true. The negated conjunction, on the other hand, adequately reflects the inductive character of an empirical statement of fact belonging to the realm of the possible.

Thus there is no justification for Cicero's sarcasm, especially when he tries to ridicule Chrysippos by amplifying and asking if a doctor should replace the conditional "if a person's pulse is agitated, he has got fever" by "not both: a person's pulse is agitated and he has no fever" or if one should say "not both: there are great circles on a sphere and they do not bisect each other" instead of the usual "great circles on a sphere bisect each other". Cicero did not grasp the real notion of Chrysippos' formulation; in the last case Chrysippos himself would have preferred the conditional form, whereas in the former, based on experience as every medical diagnosis is, the negated conjunction is better suited to the case. In connection with this it is worth while to point out that the negative formulation of empirical relations is actually the starting-point of the great laws of modern physics, as these laws are implicit expressions of experimental impossibilities, they are "statements of impotence" to overcome certain laws of Nature.[114] The negative statements with regard to the possibility of constructing a perpetuum mobile of the first or second kind or of measuring absolute simultaneity of events not coinciding in space are notable examples, throwing into relief the remarkable effort of Chrysippos to introduce such negative versions for statements of an inductive character.

[114] Cf. E. T. Whittaker, *Nature* (1942), vol. 149, p. 268.

The revolutionary progress made by Stoic physics in elaborating the concept of causality and fitting the notion of the possible into the frame of a deterministic description of the world has carried it right to the doorstep of probability, and one might have expected that the Stoics would have developed at least some of the fundamentals of the notion of the probable. Indeed, we have seen them working out clearly two logical presuppositions of this notion—the theory of disjunctive propositions and the idea of equally possible cases. Moreover, they also grappled with the third prerequisite of quantitative probability, namely the elements of combinatorial analysis.[115] Chrysippos tried to calculate the number of molecular propositions which can be formed out of all possible combinations of ten atomic propositions.[115, 116] Although the figures given are obviously wrong, the very fact that a combinatorial problem was posed and an attempted solution given indicates that the stage was set for a development which could have begun on lines similar to those which led from Tartaglia and Cardano in the sixteenth century and Pascal and Fermat in the seventeenth to Bernoulli and Laplace. Nevertheless, and in spite of the fact that throughout antiquity, as in modern times, dice and knuckle-bones were in general use in games of chance, there is no mention, even in a rudimentary form, of the most elementary quantitative concepts of probability such as the ratio of the favourable cases to all the possible ones.[117] The reasons for this failure to arrive at a concept of mathematical probability have deeper roots which we are not proposing to investigate here.[118]

[115] The first to attack a problem of this kind was Xenocrates, the third head of the Old Academy; cf. Plut., *Quaest.* conviv., 733 a.

[116] Plut., *De Stoic. repugn.*, 1047 c-d.

[117] Carneades' Theory of the probable ($\pi\iota\theta\alpha\nu\acute{o}\nu$) was a basically non-quantitative theory of reasonableness of propositions which classified them according to their degree of credibility.

[118] Cf. S. Sambursky, *On the Possible and the Probable in Ancient Greece*, Osiris XII, 35 ff., 1956.

IV

THE WHOLE AND ITS PARTS

1. *Interaction and Functional Relationship*

THE most important outcome resulting from the Stoic dynamic notion of the continuum was their doctrine of causality. However, this notion affected their physical thought in a much wider sense than that determined by an asymmetrical and time-dependent relationship such as is exhibited by the causal nexus. The Stoics, for the first time in Greek science, introduced the symmetrical concept of interaction between members of certain classes or structures. We have already seen (ch. I, §2) that hexis represents the highest type of all possible structures, i.e. an entity which exhibits "communication" (*diadosis*) between the individual members and the whole. This applied especially to the physical properties of a body which, interpenetrating one other, form its physical state. Another example of such a structure is the ethical hexis characterizing the ethical personality which embodies all the possible virtues. The interdependence[1] of virtues is one of the basic assumptions of Stoic ethics. Although every virtue is clearly distinguishable from the others, they interact in such a way that a man possessing one virtue possesses all the others. The picture that springs to the mind here, taking into account the fact that Chrysippos regarded the virtues as qualities,[2] i.e. pneumata, is of course that

---

[1] ἀντακολουθία; Sext. Emp., *Pyrrh. hyp.*, I, 68; Diog. Laert., VII, 125; Plut., *De Stoic. repugn.*, 1046 e.

[2] Galen., *De Hipp. et Plat. decr.*, VII, 2 (Arnim, III, 256, 259).

81

of total mixture, and the Neoplatonist Olympiodoros even goes back to Anaxagoras' "everything is contained in everything," the precursor of total mixture.[3] But it is of interest to notice another analogy mentioned by Clemens of Alexandria,[4] who tells of the distinction made in Stoic terminology between the asymmetrical cause-effect relation and the symmetrical relation of mutual cause and interaction.[5] Among the examples quoted are the virtues ("the virtues are each other's mutual cause in such a way that they cannot be separated because of their interdependence") together with "the stones of a vault which are each other's cause for remaining in place." Here we have the picture of an interaction of localized units of a class where the removal of one unit leads to the breaking up of the whole system held together by the laws of statics. It is very suggestive to compare these two equivalent pictures of reciprocal causality, one based on the interpenetration of pneuma tensions and the other on the equilibrium of forces acting between adjacent bodies.

Whereas in these examples the term "mutual cause" is used to describe interaction between individuals of one group, other instances quoted by Clemens show that the Stoics used "mutual cause" in a sense akin to the Aristotelian category of the "relative" and "correlative".[6] "Slave" is a relative notion, and its correlative is "master"; and the same applies to the double and its half, etc. Occasionally, Aristotle points out, the usual words do not bring out properly this correlation, as for instance in the case of "rudder" and "boat", because there are boats without rudders. He suggests the use instead of "the ruddered" and "rudder" as correlated terms. Here the characteristic difference from the Stoic approach and terminology can be seen. In accordance with their conception of causes as bodies and effect as action of one body on the other, mutual causes come into existence only during the interaction of two bodies. They are in fact the Aristotelian correlatives, or any other pair of bodies, in the state of acting upon each other. This state is always expressed by a verb, so that the last example in Stoic language would be

[3] Olympiodoros, *In Plat. Alcib.*, p. 214 (Arnim, III, 302); cf. Anaxagoras, fr. 4 (Diels, 59 B 4).

[4] Clemens Alex., *Stromat.*, VIII, 9.

[5] ἀλλήλοις αἴτιος.

[6] Arist., *Categ.*, ch. 7.

stated thus: rudder and boat are mutual causes, the rudder for moving the boat and the boat for being moved by the rudder. The analogous illustration quoted by Clemens from Stoic writings is the carving knife and the meat: "The knife is the cause for the meat to be cut, and the meat the cause for the knife to cut." Cutting and being cut are mutual causes, they are two aspects of the same event. They co-exist simultaneously; not in the Aristotelian sense of the coexistence of correlatives *qua* complementing concept,[7] but in the same way as *actio* and *reactio* come into existence or cease to exist at one and the same time. In this respect the concept of mutual causes coincides with that of the "operative cause" of the Stoic classification of causes (see p. 60).

Aristotle had already remarked that some of the relative terms can admit of variation of degree,[8] thus applying to all cases where the term can be operated in conjunction with "more" or "less." However, the dynamic notion of mutual causes and the interaction between them led the Stoics a decisive step forward in the direction of the concept of the function in cases where the causes are variable and a functional relation is thus established between variable quantities. In the case of a disease, for instance, there is functional dependence between changes in affected organs and changes of symptoms. In the source already quoted,[4] the condition of the spleen and of body temperature are mentioned as mutual causes, the swelling of the spleen and the increase of fever being regarded as interdependent quantities. As both are continuous variables, we have here the description of the change of one state as a function of another.

A few more of the extant quotations show the gradual penetration of this train of functional thought into the physical considerations of the Stoics. "The same string corresponding to its tension or relaxation, produces a high or a low pitch."[4] It is disappointing to see in this example both variables restricted to two values only, representing "opposite" states, instead of letting them pass through a continuous scale; this, however, was the heritage of Greek physics since the days of the Pythagorean "opposites." Another source, asserting as a Stoic doctrine that "the effects produced by the same cause vary along with the

---

[7] Arist., loc. cit., 7 b 15.
[8] Arist., loc. cit., 6 b 20: τὸ μᾶλλον καὶ τὸ ἧττον.

things affected and the distances,"[9] mentions the effect of the sun on the tropical, moderate and cold zones of the earth, according to its distance. The same phenomenon is described elsewhere[10] in more detail as a continuous process of dependence of the seasons on the varying positions of the sun: "Of the changes which go on in the air, they (the Stoics) describe winter as the cooling of the air above the earth due to the sun's departure to a distance from the earth; spring as the right temperature of the air consequent upon its approach to us; summer as the heating of the air above the earth when it travels to the north; while they attribute autumn to the receding of the sun from us." Here, in accordance with the Stoic conception of the corporeal nature of causes, the seasons are regarded as bodies representing air in a certain thermal state, and this state is found to be in functional dependence on the state of another body, namely the position of the sun. However, there is still a long way from these first traces of functional thought to the modern concept of the function which began to develop with Descartes (1637), and slowly crystallized out from the work of Leibniz and John Bernoulli, culminating in the mathematical definition by which there is co-ordinated to every element of a set of individuals an element of another set.

Lack of graphic representation and the failure of Greek mathematics to develop a proper algebraic notation were the main obstacles in the way of further advance of functional thinking in the wake of the first steps made by the Stoics. However, the Stoic beginnings in this field, modest as they were, were not restricted to the train of thought just described, but touched also on the problem of the general course of a function. This can be found in the discussion on the changeability of qualities in connection with the distinction made by Aristotle between two different conditions of a quality, which he defined by "habit" (hexis) and "disposition" (diathesis).[11] Habits are "more lasting and more firmly established," for example virtues "which are not easily dislodged, so as to give place to vice". Disposition, on the other hand, is "a condition that is easily changed and quickly gives place to its opposite", for

[9] Sext. Emp., *Adv. math.*, IX, 249.
[10] Diog. Laert., VII, 151.
[11] Arist., *Categ.*, ch. 8.

example heat, cold, disease, health. Habit is therefore the stronger concept and disposition the weaker one: "habits are at the same time dispositions, but dispositions are not necessarily habits." The conceptual changes of these terms exhibited in the Stoic theory are connected with the specific meaning which the term "hexis" had acquired in Stoic physics, where, as we know, it denoted the physical state of a body. A physical state can run through a whole continuum of changes, i.e. it admits of continuous variation, or, in Stoic terminology: "*hexis* can be tightened and loosened".[12] *Diathesis*, on the other hand, represents a special case of *hexis* in so far as it is "neither capable of increase nor of diminution". The example offered is the straightness of a rod, which is an exceptional state among all the possible curvatures the rod can undergo. It is a disposition in the Aristotelian sense because the slightest bending makes it vanish, which means that it is easily changeable. But the characteristic feature in the Stoic sense is that straightness is an extremal case. "Straightness cannot be loosened nor stretched nor does it admit of variation in degree." Simplicios adds to this: "They seem to understand *hexis* as the range of variation of a state and *diathesis* as the extreme case." In the same way virtues were regarded by the Stoics as *diathesis*,[13] not because of their stability, but because they represent an extremal case, a singularity.

These attempts of the Stoics to throw into relief the difference between the general course of a variable and its extremal conditions must be viewed against the background of their conception of the dynamic nature of the continuum. The point in question is not the "kinematic" definition of a geometrical figure through the movement of an element of a lower dimension, such as the generation of a line through the motion of a point. Kinematic conceptions of this kind are to be found already in pre-Aristotelian times, as proved by the passage in *De anima*: "They say a moving line generates a surface, and a moving point a line".[14] The dynamic element introduced into mathematics by the Stoics is, however, of a much more radical

[12] Simpl., *Categ.*, 237, 29 ff.

[13] loc. cit., 237, 34 f. Cf. also Stob., *Eclog.*, II, 97, 5, and Cicero, *De finibus*, III, 14, 45.

[14] Arist., *De anima*, 409 a 4. Cf. I. Newton, Introduction to *De Quadratura*, § 27.

nature and consists essentially of a physicalization of geometry by endowing geometrical figures with the elastic properties of material bodies. Another quotation from Simplicios will shed more light on this aspect of the problem:[15] "The Stoics say that (geometrical) figures exhibit tensions, and this applies also to the interval between points. On these grounds they define the straight line as a line stretched to the utmost." Thus the Stoics transposed geometrical figures into material shapes and regarded them as held together by pneuma tensions which pervade every body and define its physical state. After substitution of a cord of a given length for a geometrical line, the various shapes of the cord are compared and the straightened cord is seen as the extremal case as against the curved ones. A problem of the calculus of variations is thus stated in an intuitive way: among all curves of a given length the straight line represents the curve between whose extremities lies the greatest distance. Simplicios adds his critical comment to this procedure: "This would destroy the essence of mathematics which is static and free from every change and therefore also from tension." It is precisely the non-static approach of the Stoics that brought them a considerable step nearer to the understanding of the variable and the function, and of the elements of differential geometry.

Simplicios' criticism, made in the true Aristotelian spirit, is fully understandable. The Stoic conception represents the greatest possible departure from the static world of Aristotle, whose order was defined by distinctly separated entities of given form and sharp contours. More than that: the whole Aristotelian order was turned topsy-turvy, because now it was no longer Form that was the ultimate principle which determined and explained everything else, including motion, but motion itself was supposed to explain Form. In the "tensional geometry" of the Stoics the rigid separateness of forms was replaced by a continuum of shapes each emerging from its neighbour through a dynamic principle of variation. "But," declares Simplicios, "according to Aristotle tension is not the cause of form, because this would mean that motion and change would become the cause of quality . . . but quality is considered as having definite measures, whereas motion is an indeterminate thing and rather demands to be determined."[16] One has there-

[15] Simpl., loc. cit., 264, 34 ff.   [16] Simpl., loc. cit., 265, 5 ff.

fore, concludes Simplicios, to reject the idea, foreign to Aristotle's doctrine, that there is "a primary force which organizes the definite forms and unites them and is at once present in all of them, thus holding the group together".

If one remembers that Stoic physics matured during the golden period of Hellenistic science, that Zeno was a contemporary of Euclid, Chrysippos of Archimedes, and that Poseidonios lived one generation after Hipparchos, the question presents itself whether there are indications of an interconnection between the Stoic notions and the mathematical development of that period. As regards the physicalization of geometry, a similar trend can be observed in Archimedes' method of investigating certain mathematical problems by means of mechanics.[17] Archimedes explains the usefulness of his mechanical method, which, although it cannot replace a rigorous geometrical proof, is sufficient to supply preliminary knowledge on some questions and to suggest the truth of certain theorems in quadrature and cubature. It is further of interest to compare the Stoic "variational" definition of a straight line, put as a maximal case among lines of equal length, with that of Archimedes, who sees it as a minimum solution among all lines having the same extremities.[18] The variational principle is explicitly applied in Archimedes' second assumption, where he considers all lines in a plane having the same extremities and states that "whenever both are concave in the same direction, . . . that which is included is the lesser".

Similar influences are apparently reflected in a passage of Plutarch,[19] who raises the question of why Plato in his theory of geometrical atomism in the *Timaios* made use only of figures bounded by straight lines and completely disregarded circles. Plutarch offers reasons for "the natural priority of the straight line over the circle". His arguments are partly based on the variational approach. On the one hand, he stresses the extremal character of the straight line as the shortest connection between two points and, generalizing Euclid's definition ("a straight line is that which extends equally with the points on it"), adds that "a circle is concave with respect to its interior and convex with respect to its exterior". On the other hand, comparing a straight

---

[17] Cf. Archim., *The Method.*
[18] Archim., *On the Sphere and Cylinder*, I.　　　[19] Plut., *Plat. quaest.*, V.

line with circles of different radii, he points out that "a straight line, be it short or long, maintains its straightness whereas the curvature of circles increases with their smallness and vice versa".

Finally we have to consider briefly one more instance which illustrates the beginnings of functional thought in Stoic philosophy. Although this particular example pertains to Stoic logic, it clearly points to their tendency to replace discrete values by continuously changing variables. Among the most noted contributions of the Stoics to logic were their systematization of and additions to the theory of propositions. Among the more important molecular propositions, together with the conditional, the conjunctive, the disjunctive and the causal, we find also the comparative proposition[20]—the "indication of more" and the "indication of less".[21] The examples given are: "It is more day than night" and "It is less night than day." Here we witness the introduction of a gliding scale into the theory of propositions which in fact means that the disjunction ("it is either day or night") is generalized into a probability proposition. The alternative 1 or 0 of the two-valued proposition is replaced by a many-valued one with a gliding scale of values between 1 and 0. Actually these last two represent only the two "pure" cases forming the extremities of a continuum of "mixed" cases. It was Aristotle who, for the first time, in his classification of movements and changes, considered from an empirical point of view the mixture as an intermediate case between opposites, and discussed transitions from the intermediate to one of these opposites, for example the relation between grey on the one hand and black and white on the other.[22] But the Stoics, by establishing the mixed case as a propositional "functor", elevated the comparative proposition to a logical tool of inductive science on an equal footing with the disjunctive and conditional propositions which were so successfully developed by them.

2. *The Notion of the Infinite*

Among all the scientific subjects dealt with by the Stoics there is probably none where we have to deplore the loss of

---

[20] Diog. Laert., VII, 72.
[21] διασαφοῦν τὸ μᾶλλον, διασαφοῦν τὸ ἧττον.
[22] Arist., *Phys.*, 224 b 30 f.; 229 b 15 f.

their writings more than those pertaining to the notion of the Infinite. The few sporadic quotations, which have come to us mainly through the essays of Plutarch, give us but an inkling of their contribution to a field of thought which occupied some of the best minds of Greek science for several centuries. Nevertheless, these scanty remnants luckily preserved for us give a sufficiently clear picture of the remarkable advances made by the Stoic School in the logistics of the Infinite and of their anticipation of some of the concepts formed in modern mathematics, beginning with the calculus and leading up to the fundamentals of the theory of sets.

Greek mathematics had been confronted in a dramatical way with the problems of continuity and the infinite by the famous paradoxes of Zeno of Elea in the first half of the fifth century B.C. These paradoxes, related by Aristotle,[23] propound, under the provocative guise of arguments against the possibility of motion, some of the essential questions connected with the conception of space as a continuum of points. We shall return to this question when discussing the Stoic conception of time and the difficulties encountered when time was regarded as composed of an infinity of "nows". At the root of these difficulties and similar ones lies the failure of the ancient Greeks to master the central concepts of the infinitesimal calculus, namely the limit and the process of convergence towards the limit. Whereas Zeno's arguments touched mainly upon the theoretical aspect of the problem, the mathematicians from the fifth century B.C. onward were confronted with the intricacies of the practical side when developing methods for the computations of areas and volumes.

Antiphon the Sophist, a contemporary of Socrates and of the mathematician Hippocrates of Chios, was apparently the first who used the process of exhaustion for the computation of the area of a circle.[24] He started with an inscribed polygon, say a square, built isosceles triangles on each of its four sides, with their vertices on the circumference of the circle, then he built triangles on each side of the octagon thus defined, and continued this procedure on the sides of the sixteen-sided polygon, etc., till

---

[23] Arist., *Phys.*, VI, ch. 9.
[24] Simpl., *Phys.*, 54, 20-55, 24. The Greek term διαπανεῖν literally means "using up".

the whole area of the circle was "used up" by a polygon "whose sides, because of their smallness, coincide with the circumference of the circle".[25] Thus Antiphon seems to have held the naïve view that the process of exhaustion is finite, and similar views were still expressed in later times as shown by a passage in Plutarch[26] in a context quoted earlier: ". . . one could thus surmise that the circumference of a circle is made up by the synthesis of many small straight lines".More than four hundred years after Plutarch, however, Simplicios remarks that Antiphon's view "would annul the geometrical principle of infinite divisibility of magnitudes".[27] This was no doubt also the opinion of Antiphon's contemporary Bryson who considerably improved on Antiphon's method by intercepting the circumference of the circle between inscribed and circumscribed polygons.[28] "The circle is greater than all inscribed polygons and less than all circumscribed polygons." After this statement one would have expected some considerations of convergence with regard to the series of perimeters of the circumscribed polygons, decreasing in length, and that of the inscribed ones, with the perimeter increasing in length, but both converging toward the same limit. However, nothing of this sort follows, because Bryson, by an erroneous assumption, believed he could avoid any consideration involving infinite sequences. As the polygon "between" ·the inscribed and circumscribed polygons—which obviously means the arithmetic mean of these—also fulfils the conditions of being greater than all the inscribed ones and less than all the circumscribed ones, Bryson supposed that it must be equal to the circumference of the circle.

Antiphon's and Bryson's method was generalized by Eudoxos (first half of the fourth century B.C.), whose reasoning is perfectly correct and rigorous and avoids the mistakes of his predecessors. However, he too keeps clear of any infinitesimal procedure and bases his proof on a *reductio ad absurdum*. A well-known application of Eudoxos' method is Euclid's demonstration of the theorem that the area of circles are one to another as the squares of their diameters.[29] Eudoxos also applied the

[25] loc. cit., 55, 7.
[26] Plut., *Plat. quaest.*, 1004 b.
[27] Simpl., loc. cit., 55, 22.
[28] Themist., *Phys.*, 19, 6-17.
[29] Euclid, *Elements*, XII, 2 (cf. Heath's translation, III, 371).

method of exhaustion to the computation of volumes, and in this way he demonstrated by means of inscribed and circumscribed cylinders and prisms that the cone is a third part of the cylinder and the pyramid of the prism, which have the same base and equal height.[30] Archimedes, who refers to this fact, also made use of the exhaustion process and, with the help of inscribed and circumscribed polygons, carried out his famous approximate calculation of π.

While the method of exhaustion thus avoided by means of formal logic problems of convergence and the concept of limit, another school of thought tried to cut the Gordian knot by postulating atomic lengths. The father of this school was Xenocrates (second half of fourth century B.C.), a pupil of Plato and a teacher of Zeno the Stoic. Aristotle refers to him in the *Physics* by saying "they yielded to the argument from bisection by positing atomic magnitudes",[31] and Aristotle's commentators expressly mention his name. Simplicios[32] quotes Alexander Aphrodisiensis as saying that Xenocrates introduced the atomic lengths[33] in order to get rid of the "contradiction" implied in Zeno's dichotomy paradox that a finite length is composed of an infinity of points. The same account is given by Philoponos in his comments on the passage in Aristotle.[34] Some more information on this point is to be found in another passage of Simplicios where he comments on Aristotle's criticism of Plato's theory of geometrical atomism in the *Timaios*.[35] Here Xenocrates is mentioned together with Democritos in a context which makes it plausible to assume that Xenocrates' hypothesis was induced directly by the atoms of Democritos. It was the main attributes of the atoms—their being impassive and without quality— which led Xenocrates to the assumption of an atomic length which is "insensitive" to further division. Thus by introducing a minimum length Xenocrates tried to avoid an "infinity catastrophe" in a way reminiscent of similar attempts in modern physics. On the other hand, one can compare these lengths, which were assumed small but finite, with the intuitive and

[30] Cf. Archim., *TheMethod*, Introduction.
[31] Arist., *Phys.*, 187 a 3.
[32] Simpl., *Phys.*, 138, 3-18.
[33] ἄτομαι γραμμαί.
[34] Philoponos, *Phys.*, 83, 19 f. and 84, 15 f.
[35] Simpl., *De caelo*, 665, 5 f.

approximate conception of a differential in mathematics and theoretical physics which was occasionally used already in this sense by Leibniz. This reminds us again of the passage in Plutarch,[26] because such a differential is supposed to be of a length along which the difference between a curve and its tangent is indiscernible. However, one should not forget that the differentials, being a convenient approximate tool, have survived the rigorous considerations of limit and convergence, whereas Xenocrates' atomic lengths were introduced in order to evade the intricacies of infinitesimal calculus.

Xenocrates' indivisible segments were no doubt a welcome answer to geometrical paradoxes such as that described by Sextus[36] which is nothing else but Zeno's dichotomy problem in another formulation. If a straight line revolves round one of its ends as centre, the sum total of its points will describe an aggregate of concentric circles. If this aggregate forms a continuum, there can be no space separating one circle from its neighbour, and all the circles would thus coincide and become one. But if each circle were separated from its neighbour by a finite distance, there would be points on the line which would not participate in the revolution—which does not make sense either. The second alternative, however, would have been just the solution offered by Xenocrates' doctrine, according to which there would be a small "quantum jump" from one circle to the next corresponding to the small but finite length of the atomic elements of which the line is made up.

Sextus' paradox is a problem in two dimensions. Transposed into three dimensions, however, it remains a dilemma to which even Xenocrates' idea offers no satisfactory solution. This was already recognized by Democritos when he asserted that the volumes of the pyramid and the cone are one-third of the volumes of a prism and a cylinder respectively, having the same base and equal height—a theorem which was later proved by Eudoxos.[30] Democritos, in anticipation of Cavalieri's principle, probably conceived of a solid as being the sum of a very large number of very thin laminae and concluded that the volumes of two solids are equal if sections of each of them at the same height always give equal surfaces.[37] This course of argument,

---

[36] Sext. Emp., *Adv. math.*, IX, 418-425.
[37] Cf. T. L. Heath, Introductory note to *The Method* of Archimedes, p. 11.

together with the assumption of "atomic lengths", led Democritos to the formulation of the following paradox[38]: "If a cone were cut by a plane parallel to the base, what must we think of the surface of the sections? Are they equal or unequal? For, if they are unequal they will make the cone irregular, as having many indentations, like steps, and unevenesses; but if they are equal, the sections will be equal, and the cone will appear to have the property of the cylinder and to be made up of equal, not unequal circles, which is quite absurd."

Obviously there is no answer to Democritos' dilemma within the static concepts of atomic lengths, i.e. of *constant* magnitudes, be they assumed as small as one likes. Only the dynamic concept of the limiting process and the other notions of the infinitesimal calculus can furnish the tools for a satisfactory solution to the problem. The few fragments in Plutarch make it seem probable that it is Chrysippos to whom the credit is due of having first grasped the concept of the limit. There is of course no question whatsoever of an establishment of the existence of a limit nor even, as far as we know, of a mathematical formulation. This would have been extremely unlikely in view of the lack of any proper algebraic notation in ancient Greece. But there is no doubt as to the distinct attempt of Chrysippos to create a suitable terminology for the limiting process which breaks away from both the elusive attitude of Eudoxos and the static notions of Xenocrates.

Chrysippos' reply to Democritos is quoted by Plutarch as follows[39]: "The surfaces will neither be equal nor unequal; the bodies, however, will be unequal, since their surfaces are neither equal nor unequal." The first part of this sentence refers to the process of convergence towards the limit. Characteristically, the statement is given a form negating the law of the excluded middle. Indeed, this law is violated if the essentially static notions of equal and unequal are applied to the conception of a dynamic approach to zero. The surface of any section of the cone is either unequal to that of a given section if the distance between both is unequal to zero, or both surfaces are equal if the distance between the sections is zero—*tertium non datur*. However, if we consider the infinite series of sections which is approaching the given section when the distance

---

[38] Plut., *De comm. not.*, 1079 e.   [39] loc. cit., 1079 f.

between each member of the series and the fixed section is approaching zero, we have to discard the static concept of equal and unequal, taking into account that for each given difference in surfaces one can determine a distance which will yield a still smaller difference. This is what Chrysippos intended to express by the formula "neither equal nor unequal".

This interpretation of Chrysippos' terminology for the limiting process is suggested by yet another formulation given by him and reported by Plutarch.[40] Referring to the laminae into which a triangular pyramid is divided by parallel sections, Chrysippos says that of the sides of two adjacent ones, although being unequal, one is "greater but not exceeding".[41] This expression, later repeated again, was coined in the same deliberate attempt to describe the process of convergence towards a limit. The apparent *contradictio in adjecto* involved in it finds its explanation in the fact that the common use of "exceed" refers to a *fixed* difference between two quantities, whereas Chrysippos actually transcribes in words the relation $b > a$, where $b = a + \epsilon$ and $\lim \epsilon = 0$. In this connection it is of interest to note that the expression "greater but not exceeding" was apparently formulated in clear contradiction to "greater and exceeding", a combination often used in the comparison of different quantities, also in later literature.[42] Further, the verb "exceed" was used not only in the arithmetical sense[43] but also in the sense of "protrude", for instance in the description of the blade of a saw which is jagged, alternatively protruding and indented.[44] Of two sections of a cone, between which the distance is approaching zero, one is admittedly greater than the other, but "it does not protrude" as is the case with the fixed indentations of a saw.

How careful Chrysippos has been in maintaining the terminology of "fluxions" and avoiding any expression which could be interpreted as a reference to an *actual* infinitely small quantity —which he rejected as did the Peripatetics—is evident from another source.[45] There we are told that Chrysippos did not like the usual definition of the infinite divisibility of matter because

[40] loc. cit., 1079 d; 1080 c.
[41] μεῖζον οὐ μὴ ὑπερέχον.
[42] Cf. e.g. Simpl., *Categ.*, 162, 23.
[43] Cf. Archytas, fr. 2, Diels.
[44] Simpl., loc. cit., 263, 4.
[45] Diog. Laert., VII, 150.

of its ambiguous meaning which is thrown into relief by the literal translation "division into the infinite".[46] "There is nothing infinite into which the division leads, but the division is unceasing."[47]

There still remains the second part of Chrysippos' reply to Democritos to be dealt with: "The bodies will be unequal, since their surfaces are neither equal nor unequal". The term "body" obviously refers to the solid contained between two parallel sections of the cone. Let $A_1 < A_2 < A_3$ be the surfaces of three adjacent sections. Chrysippos' assertion is that the volume defined by the surfaces A_1 and A_2 is not equal to that defined by A_2 and A_3, in spite of the relations $\lim (A_3 - A_2) = 0$ and $\lim (A_2 - A_1) = 0$. There is no indication that this proposition has been proved by Chrysippos, and it is most unlikely that a rigorous proof was given at the time. However, the proposition is necessary in order to assure that by the limiting processes with regard to adjacent sections the volumes bounded by these sections do not become equal, which would lead to a cylinder instead of a cone and thus restore the dilemma of Democritos.

The clear conception which Chrysippos had of the limiting process removed all obstacles in the way to a deeper understanding of the nature of infinite sequences of inscribed and circumscribed figures which Greek mathematicians carefully avoided when using the method of exhaustion. More than that: the process of inscription and circumscription, used for instance by Archimedes as an approximative method for the calculation of π, was turned by the Stoics into a rigorous definition of a given body embedded in a continuum. In the dynamic continuum of the Stoics the static notion of the Aristotelian "place" could not be upheld any more. Aristotle had defined place as the innermost motionless boundary of the containing body at which it is in contact with the contained body, a kind of surface of the container of the thing.[48] The Stoic continuum, being a medium for transmission of physical actions and for exchange of forces, was incompatible with the existence of static and definite surfaces interrupting the free flux of phenomena. Causes, as we have seen, are bodies, and the notion of an incorporeal surface

---

[46] τομὴ εἰς ἄπειρον.
[47] ἀκατάληκτος. Cf. Epictet., *Dissert*. I, 17.
[48] Arist., *Phys*., 212 a 2-30.

forming the boundary between bodies acting upon each other would lead to considerable difficulties, as indeed the very idea of a static contact is entirely foreign to the dynamic concept of the interdependence of bodies and the continuous transfer of pneuma tensions throughout space.

The Stoics therefore discarded the conception of the distinct surface of a body, or generally a distinct boundary of $(n-1)$ dimensions forming the surface of a figure of n dimensions $(n = 1, 2, 3)$, and replaced it by an infinite sequence of boundaries defining the surfaces of inscribed and circumscribed figures which converge from both sides to the figure in question and thus define it as a dynamic entity. This is clearly shown in another fragment preserved by Plutarch[49]: "There is no extreme body in nature, neither first nor last, into which the size of a body terminates. But there always appears something beyond the assumed, and the body in question is thrown into the infinite and boundless." Plutarch naturally rejects this definition of a body by a "dynamic" approach and remarks that it leads to a negation of the very concept of inequality, as any unevenness at the boundaries of bodies would thus be blurred and smoothed out. Neither does he grasp the important consequence of this approach in which the static notion of contact between bodies is superseded by the dynamic notion of mixture.[50] Instead of the surfaces of two adjacent bodies touching each other, we now have a picture of the infinite sequences which envelop the surfaces, merging into one another and thus forming a narrow zone where the extremities of these bodies enter into a mixture. In this way the corporeality of the contact is maintained and any conceptual difficulty is removed which might arise from the notion of two-dimensional surfaces entering the description of physical phenomena which are taking place in three dimensions.

Part of the discussion on this point between the Stoics and the Sceptics is reflected in Sextus' writings.[51] Of special interest are the Sceptic arguments against the apprehensibility of bodies by contact based on their attempt to discard the notion of contact. These arguments amply demonstrate their failure to understand the gist of the Stoic doctrine.

The attempt of the Stoics to clarify the limiting process and

---

[49] Plut., loc. cit., 1078 e.   [50] Plut., loc. cit, 1080 e.
[51] Sect. Emp., *Pyrrh. hyp.*, III, 38-46; *Adv. math.*, III, 28-35, IX, 258-261.

the convergence of sequences was but one aspect of their approach to the problem of the analysis of the infinite. The extent of their remarkable efforts to elucidate the character of the continuum from every possible point of view becomes evident from the few fragments extant which prove that they laid the first foundations of the theory of sets. The main characteristic of the infinite set—the fact that it contains subsets which are equivalent to the whole—was known to the Stoics and formulated as follows[52]:"Man does not consist of more parts than his finger, nor the cosmos of more parts than man. For the division of bodies goes on infinitely, and among the infinities there is no greater and smaller nor generally any quantity which exceeds the other, nor cease the parts of the remainder to split up and to supply quantity out of themselves." Here the most important sentence is the first one. It seems to be a literal quotation from the writings of Chrysippos. The infinite sets "man" and "cosmos" are compared with their respective subsets "finger" and "man", and it is clearly stated that the subset is equivalent to its set in the sense defined by the modern theory of sets. This property of the infinite set was rediscovered after the Stoics by Galileo[53] who shows the equivalence of the denumerable set of natural numbers and its subset of square numbers.

The second sentence in the passage quoted above which gives a commentary to the first sentence, exhibits at first sight some similarity to a well-known fragment of Anaxagoras: "For in small there is no Least, but only a Lesser: for it is impossible that Being should Not-Be; and in Great there is always a Greater. And it is equal in number to the small, but each thing is to itself both great and small."[54] One could interpret these words of Anaxagoras also in the sense that space is at every place "inwardly infinite", as Weyl has put it. However, one should remember that Anaxagoras' words are found in the very specific context of his theory of seeds. Great and small for him were opposite qualities like hot and cold, etc., and he wanted to express the relative nature of qualities because owing to the

---

[52] Plut., loc. cit., 1079 a.
[53] Galileo, *Discorsi e dimostrazioni matematiche*, published 1638. Ed. Nazionale Vol. VIII, p. 78 f. (1933).
[54] Simpl., *Phys.*, 164, 17-20.

infinite divisibility of things one never reaches a part so small that it does not contain seeds of all the opposite qualities.[55] On the other hand, we have to regard the Stoic fragment as a general statement on the nature of infinite aggregates, the significance of which for the understanding of the continuum theory of the Stoics can hardly be overrated.

Another important fragment, expressly attributed to Chrysippos, supplements and amplifies the former one[56]: "If we are asked whether we have any parts and how many, and of what and of how many parts these consist, we will have to make a distinction. On the one hand we can posit large parts and say that we are composed of head and trunk and limbs—this was all that was asked and inquired about. On the other hand, if the question is carried further to the least parts, nothing of this kind can be assumed, but we must say that we are neither composed of such and such parts nor of so many, nor of finite or infinite ones." Thus Chrysippos clearly states that there is an essential difference between a denumerable aggregate consisting of a definite number of separable "macroscopic" elements and a continuous one. The continuum cannot be regarded as an aggregate of fixed elements, nor can it be said that it is composed of parts. Chrysippos saw it rather as a "medium of free becoming" in the sense in which the modern intuitionists like Brouwer and H. Weyl tried to define its ever incomplete and always fluid character. Again, Chrysippos uses a formula which contradicts the law of excluded middle: one must characterize the continuum as something that is composed "of neither finite nor infinite parts". We have to interpret this expression in the same way as that which we have encountered before—"neither equal nor unequal".[39] The continuum is not a static entity which can be defined as a sum of separate sub-aggregates or sequences be they finite or infinite; it is a dynamic whole which is always in the state of becoming.

3. *The Flux of Time*

Since the days of Galileo and Newton, the concept of physical time has been firmly established in the mathematical description

---

[55] Cf. Burnet, *Early Greek Philosophy*, 4th edit., p. 262 f.
[56] Plut., loc. cit., 1079 b.

of modern sciences. It was Galileo in his *Discorsi* (1638) who, in describing the movement of an accelerated body, made the revolutionary step of introducing time as a co-ordinate analogous to spatial co-ordinates and to express physical quantities (such as position and velocity) as variables depending on time. The geometrization of time and the introduction of the concept of function into physics was then followed by Newton's representation of velocity and acceleration within the conceptual frame of the infinitesimal calculus. Both these quantities are ratios— average velocity, for instance, is the ratio of the distance between two points to the time it takes the body to move from one point to the other. If this time, i.e. the difference between two instants, is made smaller and smaller, the distance traversed also diminishes, and, when carrying out the limiting process one arrives at an "ultimate ratio" defining the velocity at a given point. Generally speaking, Newton put it like this: those ultimate ratios are "limits towards which the ratios of quantities decreasing without limit always converge; and to which they approach nearer than by any given difference however small, but which they never go beyond, nor in effect attain to, till the quantities are diminished *in infinitum*".[57] Zeno of Elea's famous paradox of the arrow,[58] which negated motion because at any given moment the flying arrow is at rest at some point, springs from the failure to grasp velocity at any given point as given by this process of convergence.

There was still another shortcoming which blocked the way to a full solution of the problem even for the Stoics, who, as we have seen, were the first to form a clear conception of the infinitesimal process. It was the inability of Greek science to conceive and express in mathematical symbols even the most elementary of those constructs called physical quantities which, like velocity, are not pure numbers but form combinations of basic dimensions such as space, time or mass. In contradistinction, therefore, to the notion of time as it developed in the physics of the sixteenth century, it was not the mathematico-physical quantities, velocity and acceleration, which held a central position in the analysis of the phenomenon of "perpetual perishing" (as Locke characterized Time), but the less specific

[57] I. Newton, *Principia*, ed. F. Cajori (U.C.P. 1947), p. 39.
[58] Arist., *Phys.*, 239 b 31.

99

and less concise concept of motion, sometimes more closely specified by the opposite terms of swiftness and slowness.

However, at both levels of scientific attainment, the modern as the classical, there has always been the cognizance of the formidable difficulties which are rooted in the fact that time as a continuous extension presupposes the existence of an extensionless instant, a dividing mark within the continuum, whereas immediate awareness of perpetual change associated with time renders that concept of a pointlike "now" null and void. In Greek antiquity it was again the Stoics who, by virtue of their dynamic notion of the continuum, succeeded more than anyone else during the whole period to develop a satisfactory theory of the structure of time and to present a lucid analysis of the nature of its ultimate elements. The significance of this theory of the Stoics will become more evident when seen against the background of the attainments of their predecessors.

Many definitions of Time, such as those attributed to the Pythagorean Archytas,[59] and those of Aristotle,[60] and of the Stoics, Zeno[61] and Chrysippos,[62] contain, with certain variations, statements about its texture manifesting itself in the ordered succession of events, as well as about the mode of its measurement which is the result of its serial character. Aristotle's definition—"time is number of motion in respect of 'before' and 'after' "—expresses both the association of time with change and the possibility of enumerating this change. It is also evident from his analysis that he realized that the prerequisite for time measurement is a clock, i.e. a periodic mechanism, and that the revolution of the celestial sphere, being a regular circular motion, is the best measure of time "because the number of it is the best known".[63]

Aristotle's definition was criticized from several angles by his pupil Strato,[64] whose interesting arguments, some of which would perhaps not stand up to closer examination, probably also influenced the Stoic view on Time. Strato objected to the use of the term "number" in connection with time, as number

[59] Simpl., *Phys.*, 786, 12.
[60] Arist., *Phys.*, 219 b 1.
[61] Stob., *Eclog.*, I, 104, 7.
[62] Stob., *Eclog.*, I, 106, 5.
[63] Arist., loc. cit., 223 b 20.
[64] Simpl., loc. cit., 788, 36-790, 29.

is a discrete quantity whereas time is continuous. Zeno and Chrysippos put "interval" in place of "number", a term which fits the idea of continuity better and which also expresses a certain kinship between the spatial and the temporal dimensions, keeping in mind that time elapsed can be measured by an arc of a circle. It should perhaps be mentioned that, according to Simplicios, "interval" was used already in Archytas' definition, but it seems possible that this quotation is from a spurious source.

Strato, by defining time as "a quantity which exists in all actions", eliminated the word "motion" from his definition in order to avoid a confusion of "actions", i.e. of all kinematical aspects of physical phenomena including rest,[65] with the uniform, constant flux of time by which those aspects are supposed to be measured. An action is slow, if little happens during a long stretch of time, and vice versa. This was already indicated by Aristotle,[66] but the point is that Strato's remarks obviously induced the Stoics to introduce expressly into their definition the function of time as the measure of swiftness and slowness. Thus Zeno speaks of Time as "the interval of movement which holds the measure and standard of swiftness and slowness", and Chrysippos, amplifying on it, defines Time as "interval of movement in the sense in which it is sometimes called measure of swiftness and slowness, or the interval proper to the movement of the cosmos, and it is in Time that everything moves and exists".

It is worth while to compare Chrysippos' definition with that of Newton[67]: "Absolute, true and mathematical time, of itself, and from its own nature, flows equably without relation to anything external, and by another name is called duration; relative, apparent, and common time, is some sensible and external measure of duration by the means of motion, which is commonly used instead of true time; such as an hour, a day, a month, a year." What Chrysippos had in mind was apparently identical with Newton's relative time, but it is most improbable that he had the conception of a flux of time existing inde-

---

[65] According to Sextus Empiricus, *Adv. math.*, X, 177, there was a further definition by Strato saying that "time is the measure of all motion and rest".

[66] Arist., loc. cit., 218 b 14, 221 b 7.

[67] loc. cit., p. 6.

pendently of physical occurrences, i.e., of absolute time in the Newtonian sense. For him,[68] as for Plato,[69] "time was created with the heavens". Whether there was in Greek antiquity any notion of time flowing "without relation to anything external" seems extremely doubtful. Strato came perhaps nearest to this notion when he emphasized that "day and night and year are not Time nor part of Time but they are respectively light and darkness and the revolution of the moon and sun; Time, however, is the quantity in which these exist".[70] Galen, too, seems to have been of the same opinion, according to a tenth-century source (Ibn Abi Said) which declares: "Galen states that motion does not produce time for us; it only produces for us days, months, and years. Time, on the other hand, exists *per se*, and is not an accident consequent upon motion."[71]

The assumption that for the Stoics time was inseparably bound up with events is supported by the continuation of the passage in Stobaios quoting Chrysippos' definition.[62] There it says: "It seems that Time is to be taken in two senses, just like the earth and the sea and the void, namely in the sense of the Whole and its parts. In the same way as the void is all infinite everywhere, so time is all infinite in both directions; indeed, past and future are both infinite." Reference to the void is made here mainly as a parallel to the infinite extension of time, as the non-empty cosmos was regarded by the Stoics as a finite island within the infinite void. However, the analogy with earth and sea has to be interpreted as follows: in the same way as every part of the earth is earth and every part of the sea is sea, so every part of Time is time. An element of time, the shortest duration, exhibits the same character of an "interval of movement" as does a macroscopic stretch of time. It is not to be compared to the mathematical continuum of a line which can be whittled down to extensionless points, but is composed of substantial elements of events similar to the earth and the sea, each of whose elements are again of an earthlike and sealike character.

The last quotation bears a strong similarity to Kant's passage on intensive quantities in the *Critique of Pure Reason*, in the

[68] Philo, *De mundi opificio*, 26.
[69] Plato. *Timaios*, 38 b.
[70] Simpl., loc. cit., 790, 13-15.
[71] S. Pines, *Proc. Amer. Acad. for Jewish Research*, XXIV (1955), p. 111 f.

section on anticipation of perception[72]: "Space and Time are *quanta continua*, because there is no part of them that is not enclosed between limits (points and moments) such that this part itself is again a space or a time. Space, therefore, consists of spaces only, time of times." The idea underlying this last sentence corresponds exactly to that of Chrysippos in his simile of the earth and the sea, and nobody before the Stoics had expressed it with such clarity. The fact that points and moments are only limits and that moments cannot be synthesized into time, as Kant emphasized further, was already clear to Aristotle, who said "in so far as the 'now' is a boundary, it is not time, but an attribute of it".[73]

Here, however, we have the crux of the problem which was first attacked successfully by Chrysippos. On the one hand the "now" is supposed to be a limit of time only, a mathematical boundary which itself is not time. On the other hand this "now" coincides with the present moment, i.e. with the only event lived by me, and which, in contradistinction to the moment that has passed and the moment to come, is coupled with the immediate awareness of reality. Greek science has through all periods been vexed by this dilemma, an echo of which is to be found in the frustrating and not very helpful analysis of the Sceptics.[74]

A plausible solution seems to offer itself in the application of Xenocrates' hypothesis of atomic lengths to the temporal dimension and the assumption of the existence of indivisible atoms of time. Such an assumption would avoid the reduction of the "now" to a shadow, a mere mathematical point. This solution was apparently suggested, either by Xenocrates himself or by someone else, but it was of course rejected by the Stoics because it is contrary to the very idea of continuity that no part of a quantity can be the smallest possible one, i.e. indivisible. "The Stoics do not admit the existence of a shortest element of time, nor do they concede that the 'now' is indivisible, but that which someone might assume and think of as present is according to them partly future and partly past. Thus nothing remains of the Now, nor is there left any part of the present, but what is

---

[72] I. Kant, *Kritik der reinen Vernunft*, 1st edit., 1781, p. 169.
[73] Arist., loc. cit., 220 a 21; 222 a 10 f.
[74] Sext. Emp., *Adv. math.*, X, 193 ff.

said to exist now is partly spread over the future and partly over the past."[75] And by the same source we are told a little further on that Chrysippos "in the third, fourth and fifth book *On Parts* declares that part of the present is future and part past".[76]

At first sight it would seem as if the Stoic refutation of Xenocrates' atomic time elements is a mere playing with words. Is not the assertion that the present moment consists of a small stretch of time spread over past and future substantially the same as the Xenocratic idea of the indivisible atom of time which was suggested for the very sake of avoiding the point-like Now? In fact, we have here two completely different conceptions whose disparity again stems from the difference between the customary static notion of the continuum and the dynamic one of the Stoics. This can be elucidated by further sentences from the passage quoting Chrysippos' view on Time[62]: "He states most clearly that no time is entirely present. For the division of continua goes on indefinitely, and by this distinction time, too, is infinitely divisible; thus no time is strictly present but is defined only loosely."[77] The "loose" definition of the present obviously results from the limiting process of infinite convergence by which it is "caught" in an operation similar to that which defined a body according to the Stoic conception (cf. p. 96 and n. 49). There the surface of the body was intercepted by two infinite sequences of surfaces of inscribed and circumscribed bodies. In the case of time, the limiting process consists in an infinite approach to the mathematical Now both from the direction of the past and from the future. In this sense, "no time is entirely present" and the present is "partly spread over the future and partly over the past", because the present is given by an infinite sequence of nested time intervals shrinking towards the mathematical "now", whereby the "lower" boundaries of each interval are points of the past and the "upper" ones points of the future. In strict conformity with the dynamic conception of continua—spatial as well as temporal— the present *qua* limit of time is not sharp but forms a fringe

[75] Plut., *De comm. not.*, 1081 c.

[76] loc. cit., 1081 f.

[77] Cf. Arios Didymos, fr. 26, where the same view is attributed to Poseidonios.

covering the immediate past and future. In contradistinction to the static concept of an "atom of time" we have thus to regard the Stoic present as a shrinking duration of only indistinctly defined boundaries. The physical significance of such a duration is that it still represents an eventlike structure, it is an elementary event, and macroscopic time is composed of the succession of such events in the same sense as every part of the earth is earth and every part of the sea is sea. The mathematical "now" towards which the shrinking intervals of duration converge has no physical significance. This "now" of Aristotle and Strato is "fleeting and next to nothing", it is "unreal and exists only in pure thought".[78]

One is tempted to attribute to Plato an anticipation of the Stoic conception of Time, when he says in the *Timaios:* "More-over, when we say that what has become has become and what is becoming is becoming, and that what will become will become, and that what is not is not—all these are inaccurate modes of expression."[79] However, this remark is too aphoristic to allow for a conclusive interpretation. On the other hand, there is a remarkable similarity of the Stoic doctrine to the ideas on time of some modern philosophers, and especially noticeable in this respect is the theory of Whitehead,[80] as some quotations from his works will prove: "A moment has no temporal extension, and is in this respect to be contrasted with a duration which has such extension. . . ." "A moment is a limit to which we approach as we confine attention to durations of minimum extension." "There is no such thing [as the instantaneous present] to be found in nature. As an ultimate fact it is a nonentity. What is immediate for sense-awareness is a duration. Now a duration has within itself a past and a future; and the temporal breadths of the immediate durations of sense-awareness are very indeterminate and dependent on the individual per-cipient. . . . The passage of nature leaves nothing between the past and the future. What we perceive as present is the vivid fringe of memory tinged with anticipation. . . . The past and the future meet and mingle in the ill-defined present."

[78] Proclos, *In Plat. Tim.*, 271 d.
[79] Plato, *Timaios*, 38 b.
[80] A. N. Whitehead, *The Concept of Nature* (C.U.P. 1920), ch. III; *Science and the Modern World* (Macmillan 1925), ch. VII; *Process and Reality* (Macmillan 1929), ch. II.

The striking close kinship of Whitehead's doctrine with that of
the Stoics could be proved by many more quotations. It lends
support to the assumption that, like Whitehead, the Stoics
made nature prior to time and identified the flux of time with the
passage of events of nature. The conception that the world is
made up of events or of "drops of experience" is inconsistent
with that of an "absolute time", and no doubt the Stoics would
have agreed with Whitehead's statement that "there is time
because there are happenings and apart from happenings there
is nothing". More specifically, the character of these happenings
and therefore that of time on a macroscopic scale reveals itself
as essentially cyclic and periodic. In this sense one has to interpret
the recurrent allusion to the "movement of the cosmos" in
Stoic definitions of time. In addition to the daily and yearly
cycles of this movement there is the cosmic period of the Great
Year. Thus the Stoic Apollodoros of Seleuca (end of second
century B.C.) said [81]: "Time is the interval of movement of the
cosmos . . . and the whole time is passing just as we say that the
year passes, on a larger circuit." This "larger circuit" is obviously
nothing else but the greatest of the cosmic periods, the Great
Year whose length the Stoics reckoned from one state of the
universe to the next identical one recurring after the world has
passed through the stage of *ekpyrosis*.[82] The idea of the Great
Year has oriental precursors and is found also in Heracleitos and
Plato.[83] The Stoics, however, were the first to identify it with
their hypothetical full cycle of cosmic transmutation of matter
which consists of the preponderance of fire in the beginning, the
differentiation of the other three elements in the intermediate
stage, and the ascendancy of fire at the end, after which the cycle
of transmutation begins anew.[84]

The essential point in our context is that for the Stoics the
length of the cycle was determined by the return of the actual
material state of the Whole, whereby each state, given by the

[81] Stob., *Eclog.*, I, 105, 8.
[82] The usual translation of ekpyrosis as "conflagration" is misleading,
because it suggests a sudden catastrophe. In fact, ekpyrosis originally denoted
that period of the cosmic cycle where the preponderance of the fiery element
reaches its maximum.
[83] Cf. J. Burnet, *Early Greek Philosophy*, p. 156 ff.
[84] Euseb., *Praep. evang.*, XV, 18; Nemes., *De nat. hom.*, ch. 38 *et alii*. Cf.
also *The Physical World of the Greeks*, p. 198 ff.

sum of all hexeis, follows from the former through a continuous transition. We have seen already an indication of this conception of the dynamic evolution of a cycle in the instance given for the comparative proposition "rather day than night" (p. 96), where different states of the daily period were regarded as "mixed cases". The same approach was found in the description of the yearly cycle (p. 91), where the different seasons and their transitions were defined in terms of different thermal states of the air, which are dependent on the position of the sun. In cosmic dimensions it is the continuous change in the mixture of the elements on a universal scale which fixes the cosmic period, and, in Stoic terminology, this change is given by the variation of the composition of the pneuma, taken over the cosmos as a whole. In Chapter I (p. 10) we have seen that *psyche* is pneuma in its driest and hottest state, whereas *physis* and the inorganic state are characterized by an increasing admixture of the humid and the cold. We have to apply this differentiation to the cosmos as an entity regarded by the Stoics as a living rational being endowed with *psyche* and *nous*.[85] The cosmic cycle thus means that the cosmos, although subject to continuous metabolism, never dies and that its immortality is only another expression of the infinite extension of time, of the never-ceasing succession of events.

Luckily, Plutarch has preserved for us two fragments from Chrysippos' book *On Providence* which give a good illustration of the Stoic conception of the eternal cosmic metabolism. "As death is the separation of soul from body[86] and as the soul of the cosmos does not leave it but is growing continuously until it has consumed all matter, one cannot say that the cosmos is mortal."[87] And further: "When the cosmos is completely in the fiery state, so at the same time are its soul and its *hegemonikon*. But if that what is left over of the soul is changing into the humid state, the cosmos is in a certain way transformed into body and soul, and thus, composed of both, it exhibits another order."[88] Plutarch, in continuing this passage, says that when *ekpyrosis*, i.e. the thermal part of the cycle, has reached its maximum, the

---

[85] Diog. Laert, VII, 138-143.
[86] Cf. Nemes., *De nat. hom.*, ch. II.
[87] Plut., *De Stoic. repugn.*, 1052 c.
[88] loc. cit., 1053 b.

cosmos is wholly soul-like, whereas in the cooling down part of the cycle, i.e. the early part of the recurring creation, the cosmic soul slackens (because of the reduced tension of the pneuma) and becomes more humid, which means that the cosmos becomes body-like. Generally, therefore, the cosmos is in a mixed state, being both body and soul, and its eternal life is characterized by periods exhibiting different ratios of this mixture and changing continuously in a slow process of dynamic transformation. However, we shall not enter here into a more detailed discussion of the doctrine of ekpyrosis as we are concerned solely with the analysis of the Stoic theory of time.

4. *The Cosmos in relation to its Parts*

Without going further into details of Stoic cosmogony we have to discuss certain aspects of their views on the structure of the universe as a whole, as their cosmological theories are closely bound up with the essential notions of their physics—pneuma, continuous forces and the coherence of a structure and the interrelatedness of its parts.

The main feature which the Stoic cosmos has in common with that of Aristotle is the geocentric order—the earth in the centre of a finite cosmos and the heavens with moon, sun, the planets and the fixed stars surrounding it. The striking difference between the two cosmologies is the Stoic conception of an infinite void extending beyond the finite cosmos. This gave rise to a host of problems, some of them similar to those confronting cosmologists of the modern era. On the other hand it is important to realize how fundamentally the approach to problems of this kind was affected by the basic physical concepts of the Stoics.

A very good example in this respect is the explanation of the state of rest of the earth in the centre of the cosmos which in earlier Greek cosmologies was usually based on considerations of symmetry. Anaximander said that "the earth keeps its place because of its indifference",[89] because by its central position it is "indifferently related to every extreme point". The application of the principle of sufficient reason in this special form was taken over by Parmenides and Democritos,[90] who described

[89] Arist., *De caelo*, 295 b 10; Hippol., *Refut.*, I, 6, 5 (Diels, 12 A 11).
[90] Aet., III, 15, 7.

the effect of the equal distance of the earth from the surrounding heavens as "equilibrium", and a similar view is found in Plato.[91] Anaximenes, who assumed the earth to be flat, explained its state of rest as resulting from the resistance offered to a flat body by the air underneath it and thus introduced force as the cause of rest.[92]

In Stoic cosmology, geometrical symmetry as cause of rest was replaced by the picture of a symmetrical action of forces upon the spherically shaped earth. We are told of two analogies which served as illustrations of this action: "[The Stoics] use the following example to prove the state of rest of the earth. If one throws a grain or the seed of a lentil into a bladder and blows it up by filling it with air, the seed will be raised and stay in the middle of the bladder. In the same way, the earth will remain staying in the centre, being kept in equilibrium by the pressure of the air from all sides. And again, if one takes a body and ties it from all sides with cords and pulls them with precisely equal force, the body will stay and remain in its place, because it is dragged equally from all sides. . . ."[93] It is interesting to note that the Stoics felt compelled to add the second analogy to the first one, although the idea was to illustrate the result of symmetrical pressure. The reason is that the second case is a correct example of a dynamic equilibrium resulting from the vectorial sum of all the pulls exerted on the body being zero. The first case, of course, does not supply the precise answer required, as it will show only that the grain is raised and whirled about by the stream of air. As far as the hypothesis itself is concerned, it is probable that air is used here synonymously with pneuma, whose action being similar to that of a symmetrical force was regarded by the Stoics as a cause of the state of rest of the earth.

The function of the field of force, which not only keeps the earth in equilibrium but also preserves the unity of the cosmos as a whole and maintains its stationary state, held a central place in the cosmological discussions in Stoic writings. The various passages quoted by Plutarch in his polemic against the Stoics are somewhat contradictory and do not give a clear picture, but it would certainly be unfair to judge the consistency of the

[91] Plato, *Phaedo*, 108 e.
[92] Arist., loc. cit., 294 b 15.
[93] Achill., *Isagoge*, 4 (Arnim, II, 555).

Stoic theory by these fragmentary quotations out of context. Still, a number of important conclusions can be drawn from these as well as from other sources. The theory of the finite cosmos embedded in the infinite void is clearly expounded in Cleomedes' *"On the circular motion of the celestial bodies"*, the treatise of a Stoic writer of uncertain date (probably first century A.D.)[94] Here the existence of a void as a receptacle for the cosmos is postulated against the Aristotelian view and it is shown further that, taken the void as given, it cannot be finite but must have infinite extension.[95] The existence of the void allows also for the variations of the volume of the cosmos which expands during the thermal period of the cosmic cycle and later contracts again.[96]

The idea of a vacuum within the cosmos itself is emphatically rejected,[97] because this would destroy the coherence of the universe and interrupt the sympathy of its parts, which are maintained by the pneuma which continuously pervades it (cf. ch. II, p. 41). We know already that the Stoics looked on the cosmos as an organic entity, and the cosmic sympathy was only another expression for the hexis holding this entity together. But it is quite understandable that they made an attempt to explain this sympathy on more specific grounds, and to base the explanation on the material composition of the universe as a whole and on structural considerations. In connection with this, it is of interest to know what a later source already quoted (about third century A.D.) relates in the name of Chrysippos.[98] The reason for the state of rest of the universe, it says, is the "equal weight" (*isobares*), or the equilibrium of the four elements. The cosmic order is the result of the mixture of the two heavy elements with the two light ones. Thus, if the cosmos were heavy, it would move downwards, if light—upwards. "It remains in a state of rest because heavy and light have an equal share." This picture of a dynamic equilibrium of the cosmos within the void is no doubt an offspring of the old Heracleitean doctrine of the upward and downward motion of fire and water which regulates the world order. It is worth while mentioning

[94] Cleom., *De motu circulari corporum caelestium*, ed. Ziegler, Leipzig, 1891.
[95] loc. cit., I, 1, p. 4 f., p. 12 f.
[96] loc. cit, p. 16.
[97] loc. cit., p. 8.
[98] Achill., *Isagoge*, loc. cit.

that Cleomedes also quotes Heracleitos when talking about the exchange between heaven and earth[99]: "The way upwards and downwards, says Heracleitos, goes through the whole of the material world which is in constant change and which submits in everything to the Demiurge for the administration and permanence of the universe." On the other hand, it is not unlikely that the picture of the universe floating in the void was used by Chrysippos in analogy with the hydrostatic propositions of his contemporary Archimedes. The term *isobares* (of equal weight) applied by Chrysippos according to the quotation in Achilles appears also in the third proposition of Archimedes' first book *On Floating Bodies*. It states that a solid which, volume for volume, is of equal weight with the fluid in which it is immersed, will neither sink nor rise. Similarly, Chrysippos might have had in mind that the cosmos is mixed of heavy and light elements in a proportion which balances their opposite tendencies and keeps the whole together.

There seem to have been other theories propounded by Chrysippos and other Stoic authors explaining the coherence of the universe against the external void and its state of rest. The spherical shape of the cosmos suggested some kind of universal attraction of all its parts towards the centre. "Spherical structures necessarily tend from their surfaces towards their centre and thus have a direction downwards, towards which they tend. This applies also to the cosmos which has a spherical shape and where 'below' and 'centre' coincide."[100] This conception, reminiscent of gravitation, would contradict the idea of absolute "lightness" attributed to air and fire and make them only light in relation to earth and water. Indeed, among the Chrysippean fragments quoted by Plutarch there is one defining fire negatively as "not heavy", and another stating that "air in itself possesses neither gravity nor lightness".[101] Plutarch then quotes passages that refer to the natural tendency of the cosmos "towards the centre" and which according to him are proof of the utter absurdity of the Stoic theory, as by his interpretation "centre" refers to the infinite void, and it goes without saying that with respect to the void no geometrical location whatsoever makes

[99] Cleom., loc. cit., I, 11, p. 112.
[100] Cleom., loc. cit., I, 1, p. 18.
[101] Plut., *De Stoic. repugn.*, 1053 e; cf. ch. I, p. 6.

any sense. However, Plutarch obviously misinterpreted Chrysippos' somewhat ambiguous formulation in one of the passages where he talks of the stability of the cosmos "secured by the spatial occupation in the centre which would be upset if it were located elsewhere".[102] This and an immediately following quotation stating that "matter eternally occupies the central place" actually refer to the spherical shape of the cosmos, i.e. to the symmetrical grouping of the material world round a central point.

This interpretation is supported by other, more clearly formulated passages from Chrysippos' writings such as the following where he apparently refers to Aristotle's characterization of empty space as being devoid of any distinct direction,[103] and concludes from this that space itself cannot be the reason for the agglomeration of the material world in one place: "The void admits of no difference by which bodies would move rather here than there; but it is the constitution of the cosmos that is the cause of their tendency and movement from all sides towards the centre."[104] Thus the material world contains in itself the reason for its organization into a coherent entity. "It is its nature to move together with all its parts towards its union and cohesion, and not towards its dissolution and dispersion."[105] And further: "It is plausible that all bodies have a primary natural movement towards the centre of the cosmos."[106] And still more generally: "The material world, in its movements towards the centre and away from the centre, is controlled and held together."[107] In addition to the attracting cosmic force which obviously is the cause of this natural movement, Chrysippos also supposed the existence of some kind of inertial forces which would sustain the central motion even if the connection between one part and the rest of the cosmos were interrupted. "However each of the parts, naturally attached to the others, might move, it stands to reason that it would move thus also by itself, if for the sake of argument we assume it to be in some void within the cosmos. For in the same way as it moves towards the centre, being

---

[102] loc. cit., 1054 c.
[103] Arist., *Phys.*, 214 b 30 ff.
[104] Plut., loc. cit., 1054 d.
[105] Plut., loc. cit., 1054 f.
[106] loc. cit., 1055 a.
[107] Plut., *De def. orac.*, 425 d.

held together from all sides, it would persevere in this motion even if for instance there suddenly existed a void round it."[108]

Thus it appears from all these fragmentary quotations that the Stoics had a definite conception of some cosmic dynamical law by which the hexis of the universe within the void was maintained. It was this notion of cosmic sympathy which Cleomedes used for instance to refute the Peripatetic arguments against the Stoic cosmos: "They also argue that if there existed a void outside the cosmos, matter would be poured out into it and dispersed and scattered into infinity. Our reply is that this could never happen, because the cosmos has a hexis which holds it together and protects it, and the surrounding void can not affect it. It maintains itself by the rule of an immense force, contracting and again expanding into the void according to its natural transmutations, alternately dissolving into fire and starting creation again."[109]

However, the main point of the later Peripatetic criticism was not the dissipating effect of the void. They were ready to concede that—from the point of view of the Stoics—hexis would prevent such dissipation. Alexander Aphrodisiensis, according to a quotation in Simplicios,[110] expressed this opinion, but added that hexis, although responsible for the coherence of the universe, could not make it remain at rest and prevent its motion through the infinite void. Here Alexander seems to have misinterpreted Aristotle's argument in his rejection of the void—since Aristotle expressly stated that things in the void must be at rest,[111] or else Alexander must have been influenced by Epicurean kinematics. What makes this argument between the two schools so interesting is that while the Stoics were in complete agreement with Aristotle as regards the negation of any void *within* the cosmos, they based their view solely on the necessity of the continuous extension of the pneuma. On the other hand, the fact that they emphasized the state of rest of the cosmos while emphatically insisting on the existence of the void as the receptacle of the whole shows that the Epicurean kinematics of bodies falling freely through the infinite void carried no weight with them.

---

[108] Plut., *De Stoic. repugn.*, 1055 b, c.
[109] Cleom., loc. cit., I, 1, p. 10.
[110] Simpl., *Phys.*, 671, 4-15; cf. also Themist., *Phys.*, 130, 13-17.
[111] Arist., *Phys.*, 214 b 31.

It is important to emphasize again that the Stoics, although they saw the cosmos embedded in an infinite void, regarded it as a closed system unaffected by the vacuum outside it. In spite of the fact that they used a special term to denote the sum total of the cosmos and the void, calling it The All,[112] they did not attribute to the void any corporeal properties whatsoever, and there was therefore no question of looking at the material universe as a partial system of the All. We have to bear this in mind when we discuss now some Stoic fragments concerning the character of the cosmos as a closed system. One of these state that "the cosmos is a perfect body but its parts are not perfect, as they exist only relative to the whole and have no absolute existence".[113] The term "relative" corresponds here exactly to the fourth Stoic category, meaning that every partial system of the cosmos must be regarded as having some relation to the cosmos as a whole, and it is in this sense only that the cosmos in its totality can be denoted as a perfect body.

This view is brought out still clearer in another passage: "Of the cosmos alone can it be said that it is self-supporting (*autarkes*), because it alone contains within itself all that it needs. It is fed and grows out of itself, whereas its parts are in mutual exchange with each other."[114] This mutual exchange means mutual dependence: one partial system in the course of the life of the cosmos might gain at the expense of another, and vice versa; the cosmos as a whole, however, being the sum total of all possible partial systems, is perfect in the sense of independence and self-support. Here the Stoics hit upon an important physical law which applies to closed systems that are not subject to any interference. All the forces acting in such a system are inner forces and their sum total vanishes. Any partial system, being open to influences from other parts, is imperfect in the sense that no such conservation law can be formulated for it and that changes occurring in it are the result of forces which under the prevailing circumstances are external and imposed from the outside.

Seen from the partial, i.e. the non-closed system, the external forces will thus appear as opposed to its own course, and only

---

[112] Aet., II, 1, 7.
[113] Plut., loc. cit., 1054 f.
[114] loc. cit., 1052 d.

when the other partial systems from which these forces originate are included and integrated into one single system, does this opposition vanish and the motions of all the parts appear as one natural co-ordinated entity. Chrysippos, according to Plutarch, again and again expressed this principle in his physical writings and stated: "There are many obstacles and impediments for partial entities and movements, but none for the whole."[115]

In quoting this sentence we have touched upon the borderland between Stoic physics and ethics. Acting in accordance with the moral law meant for the Stoics acting from a point of view which takes into account "universal nature" (*koine physis*), i.e. the endeavours of the individual as an organic part of the endeavours of society as a whole.[116] Four hundred years after Chrysippos, Marcus Aurelius, alluding to the sentence just quoted—and repeating the term for obstacle which is used here[117]—translated this physical statement into a maxim for the moral attitude of the individual and its actions[118]: "Did any obstacle oppose you in your effort towards an object? If indeed you were making this effort without any reservation, this obstacle is at once injurious to you as a reasonable being. But if you take into consideration the common lot,[119] you are not yet hurt nor hindered." One can hardly find a more striking example of the well-known Stoic philosophy of life which founded basic principles of ethics upon those of physics.

[115] Plut., loc. cit., 1056 e.
[116] The influence of Heracleitos on this philosophy can be recognized for instance in the fragment quoted by Sextus Empiricus, VII, 133.
[117] ἐνστήμα..
[118] Marc. Aurel., *Medit.*, VIII, 41. Cf. also loc. cit., II, 9.
[119] τὸ κοινόν.

APPENDIX:
TRANSLATIONS OF TEXTS

For part of the sources quoted in the following, available English translations have been used (with minor changes), as indicated at the end of these passages. All other texts are in the author's own translation.

Chapter I

Alcmaeon of Croton, fragment 4, Diels (from Aetios, V, 30, 1)

Health is the equality of rights of the functions, wet-dry, cold-hot, bitter-sweet and the rest; but single rule among them causes disease; the single rule of either pair is deleterious. . . . Health is the harmonious mixture of the qualities. (K. Freeman.)

Hippocratic School: *Ancient medicine*, XVI

And I believe that of all functions none hold less power in the body than cold and heat. My reasons are these: so long as the hot and cold in the body are mixed up together, they cause no pain. For the hot is tempered and moderated by the cold, and cold by the hot. But when either is entirely separated from the other, then it causes pain. (W. H. S. Jones.)

Diogenes of Apollonia, fragment 5, Diels (from Simplicios, *In Phys.*, 152, 22-153, 13)

And it seems to me that which has Intelligence is that which is called Air by mankind, and further, that by this, all creatures are guided, and that it rules everything; for this in itself seems to me to be God and to reach everywhere and to arrange everything and to be in everything. And there is nothing which has no share of it; but the share of each thing is not the same as that of any other, but on the contrary there are many forms both of the Air itself and of Intelligence; for it is manifold in form: hotter and colder and drier and wetter and more stationary or having swifter motion; and there are many other modifications inherent in it and infinite variations of savour and colour. Also in all animals the soul is the same thing—Air, warmer than that outside in which we are, but much colder than that nearer the sun. This degree of warmth is not the same in any of the animals (and indeed, it is not the same among different human beings), but it differs, not greatly, but so as to be similar. But in fact, no one thing among things subject to modifications can possibly be exactly like any other thing, without becoming the same thing. Since, therefore, modification is manifold, animals are also manifold and many and not like one another either in form or in way of life or in intelligence, because of the large number of modifications. Nevertheless, all things live, see and hear the same thing, and all have the rest of Intelligence also from the same. (K. Freeman.)

Cicero, *De natura deorum*, II, 23-28

It is a law of Nature that all things capable of nurture and growth contain within them a supply of heat without which their nurture and growth would not be possible; for everything of a hot, fiery nature supplies its own source of motion and activity; but that which is nourished and grows possesses a definite and uniform motion; and as long as this motion remains with us, so long sensation and life remain. . . . Every living thing therefore, whether animal or plant, owes its vitality to the heat contained within it. From this it must be inferred that this element of heat possesses in itself a vital force that pervades the whole world.

We shall discern the truth of this more readily from a more detailed account of this all-permeating fiery element as a whole. All the parts of the world are supported and sustained by heat. This can be perceived first of all in the element of earth. We see fire produced by striking or rubbing stones together; and when newly dug, the earth doth steam with warmth. . . . That water also contains an admixture of heat is shown first of all by its liquid nature; water would neither be frozen into ice nor congealed into snow and hoar-frost unless it could also become fluid when liquified and thawed by the admixture of heat. . . . Indeed the air itself, though by nature the coldest of the elements, is by no means entirely devoid of heat; indeed it contains even a considerable admixture of heat, for it is itself generated by exhalation from water, since air must be deemed to be a sort of vaporized water, and this vaporization is caused by the motion of heat contained in the water. We may see an example of the same process when water is made to boil by placing fire beneath it.

There remains the fourth element: this is itself, by nature, glowing hot throughout and also imparts warmth of health and life to all other substances. Hence from the fact that all parts of the world are sustained by heat the inference follows that the world itself owes its continued preservation for so long a time to the same or a similar substance. (H. Rackham.)

Sextus Empiricus, *Adv. math.*, IX, 78-80

Some of the bodies are unified, some formed of parts conjoined, some of separate parts. Unified bodies are such as are controlled by a single hexis similar to that of plants and animals; those formed of conjoined parts are such as are composed of adjacent elements which tend to combine into one main structure, like cables and turrets and ships; those formed of separate parts are compounded of things which are disjoined and isolated and existing by themselves, like armies and flocks and choruses.

Seeing, then, that the universe also is a body, it is either unified or of conjoined or separate parts. But it is neither of conjoined nor of separate parts, as we prove from the sympathies it exhibits. For in accordance with the waxings and wanings of the moon many sea and land animals wane and wax, and ebb-tides occur in some parts of the sea. And in the same way, too, in accordance with certain risings and settings of the stars alterations in the surrounding atmosphere and all varieties of change in the air take place, sometimes for the better, but sometimes fraught with pestilence. And from these facts it is obvious that the universe is a unified body. For in the case of bodies formed from conjoined or separate elements the parts do not sympathize with one another, since if all soldiers, say, in an army have perished the one who survives is not seen to suffer at all through transmission; but in the case of unified bodies there exists a certain sympathy, since, when the finger is cut, the whole body shares in its condition. So then, the universe also is a unified body. (R. G. Bury.)

Achilles, *Isagoge*, 14 (Arnim, II, 368)

Bodies are called unified if they are governed by a single hexis, such as stone and wood, whereby hexis is the cohesive pneuma of the body. Bodies are conjoined which are not bound by a single hexis, such as a ship or a house, for the first is composed of many planks and the latter of many stones. There are discontinuous bodies such as a chorus. Of this type there are two variations, one consisting of definite and denumerable bodies, and the other of undefined ones such as a crowd.

Clemens Alexandrinus, *Stromateis*, V, 8.

What is puzzling is not the union of all things and the revolution of the cosmos according to the poet Aratos, but perhaps the tension of the pneuma which pervades and holds the cosmos together. It is better to take the aether as the all unifying and binding substance, as was also said by Empedocles.

Plutarch, *De comm. notitiis*, 1085 c

As the Stoics call the four bodies Earth, Water, Air and Fire primary elements, I do not understand how they make part of them simple and pure and part of them compounded and mixed. For they teach that earth and water are neither themselves cohesive nor make others so, but that they conserve their unity by sharing the pneuma-like and fiery force. Air and fire, however, preserve their own tension through their elasticity and, mixed with the other two, give them tension and permanence and substantiality. How could earth or water be elements, if they are neither simple nor primary nor supported by themselves but always in need of something external to hold them together in existence and to preserve them?

Galen, *Introductio s. medic.*, 9

According to Athenaios[1] the elements of man are not the four primary bodies, fire, air, water and earth, but their qualities, the warm, cold, dry and humid, two of which—warm and cold—he assumes to be active causes and two—dry and humid—passive. He introduces a fifth element in accordance with the Stoics, the all pervading pneuma by which all things are held together and controlled.

Galen, *De multitudine*, 3 (Arnim, II, 439)

The doctrine of being active towards itself or operating towards itself does not make sense, and the same applies to the faculty of

---

[1] First half of first century A.D., founder of the "Pneumatic School" in medicine.

holding itself together. And those who have most of all introduced the cohesive faculty, namely the Stoics, make one thing
cause cohesion and the other cohesive. The pneuma-like substance makes for cohesion and the hyle-like is made cohesive.
Therefore they say that air and fire make cohesive and that earth
and water are made cohesive. . . .

Neither the followers of Herophilos nor the later Stoics give any
proof for their assertion that pneuma and fire make themselves
and other substances cohesive, whereas water and earth need the
others for their cohesion. From the obvious appearance of things
one should conclude that the hard and resistent and dense will
hold themselves together, while the loose and soft and yielding
will need another substance for their cohesion. Not only do these
people expect that one will believe their hypothesis while they
furnish no evidence for it, but they do not realize that their
assumption is self-contradictory. For they say fire and air, the
most refined and soft and yielding of all substances, are the cause
of the hardness and resistance of earth, which means that something could give others a share in force or natural form or
activity, or quality, in which it itself has no part whatsoever.
Moreover, it seems obvious that not only is nothing held together
by fire but that everything is dissolved by it.

Alexander Aphrodisiensis, *De mixtione*, p. 223, 6 ff.

Is it not paradoxical to say that the whole nature is united by
a pneuma pervading all of it, and by which the universe is being
held and kept together and is in sympathy with itself? Not
knowing the foremost cause of the unity of the universe[1] . . .
and through a preconceived opinion based on several doctrines
they attribute its unity to certain bonds and material causes and
a certain pneuma pervading the whole nature. . . . If this were
really true, then whatever cohesion would be created by the
pneuma should exist in all bodies. But this is not the case, for
some bodies are cohesive and others discontinuous. It is therefore
more reasonable to say that each of them is held together and
made a unity towards itself by its own essence, in so far as it has
existence of its own, and that the mutual sympathy of the bodies
is preserved by the communion of matter and the nature of the

---

[1] The Aristotelian aether.

divine body surrounding the universe, rather than by the bond of the pneuma.

And what is the tension of the pneuma by which all parts are bound together and have cohesion in themselves and adhere to the adjacent bodies?[1] For pneuma is forcibly carried away by others because of their natural advantage over it, as it, being passive, can offer no resistance to a moving body. It can gather some strength through incessant motion, but it is passive by its own nature. It is humid and easily divisble and therefore other bodies with which it is mixed can most readily be divided. For this reason some people have believed it to be akin to the void and of insipid nature and others to contain many empty spaces. Furthermore, if the bodies are not falling to pieces but keep together because of the pneuma which makes them cohesive, then obviously those bodies which fall apart would contain no binding pneuma. . . .

Furthermore, if the pneuma, consisting of fire and air roams through all bodies and mixes with all of them, and the existence of each depends on that, how could it still be a simple body? If pneuma is composed of simple components and therefore a secondary substance, how can fire and air exist, by the mixture of which pneuma is generated—that pneuma without which no body can exist?

Alexander Aphrodisiensis, *De mixtione*, 216m 14 ff.

Chrysippos' theory of mixture is as follows: he assumes that the whole material world is unified by a pneuma which wholly pervades it, and by which the universe is made coherent and kept together and is made intercommunicating. And of the compounded bodies some become mixed by juxtaposition when two or more substances are put togther in the same place and placed side by side, joining each other, as he says, and preserving in this juxtaposition their proper essence and quality according to their individuality, as happens for instance when beans and grains of wheat get mixed with one another. Some bodies are destroyed together through a complete fusion of their substances and their respective qualities as is the case with drugs whose components undergo simultaneous destruction and as a result of it another body emerges.

---

[1] Here I read with v. Arnim δεσμῷ. τίς γὰρ . . .

Certain mixtures, he says, result in a total interpenetration of substances and their qualities, the original substances and qualities being preserved in this mixture; this he calls specifically *krasis* of the mixed components. It is characteristic of the mixed substances that they can again be separated, which is only possible if the components preserve their properties in the mixture. . . . This interpenetration of the components he assumes to happen in that the substances mixed together interpenetrate each other such that there is not a particle among them that does not contain a share of all the rest. If this were not the case, the result would not be *krasis* but juxtaposition. The supporters of this theory adduce evidence for their belief from the fact that many bodies preserve their qualities whether they are present in smaller or in larger quantities, as can be seen in the case of frankincense. When burnt, it becomes greatly rarefied, but for all that preserves its quality. Further, there are many substances, which, when assisted by others, expand to an extent which they could not do by themselves. Gold for instance, when mixed with certain drugs, can be spread and rarefied to an extent which is not possible if it is beaten out by itself. Similarly, there are cases where we can be effective together with others, while we cannot when we are alone. We can cross rivers if we form a chain, a thing we could not do by ourselves; and together with others we can bear weights which we could not bear if the whole load were to fall on us alone. And the tendrils of grape-vine which could not stand up by themselves can do so if they are entangled with each other. Thus, he says, we should not be surprised that certain substances assist each other by forming a complete union such as to preserve their own qualities while totally interpermeating each other, even if the mass of one is so slight that by itself it could not preserve its quality if spread to such an extent. Thus a ladle of wine mixes with a large amount of water, being assisted by the latter to spread throughout a great volume. In order to prove this assertion, the Stoics adduce as clear evidence the fact that the soul has substantiality of its own as has the body containing it. By totally pervading the body it preserves in this mixture its own essence (there is no part of the living body which does not have its share in the soul), and the same holds for the nature of plants and for the physical structure of those things held together by hexis.

Stobaios, *Eclog.*, I, 154

They call *krasis* the total interpenetration of two or more liquids whereby their respective qualities are preserved; for the quality of each of the mixed liquids is displayed at the same time in the mixture, e.g. wine, honey, water, vinegar, etc. That in such a mixture the qualities of the components persist is evident from the fact that they often can be separated by a contrivance. If, for example, one puts an oiled sponge into wine mixed with water, the water separates from the wine by returning into the sponge.

Simplicios, *In categ.*, 165, 32-166, 29.

The Stoics enumerate two different classes instead of one, by classifying some concepts under the relatives and some under the relative states. The relatives they distinguish from things having absolute meaning, and the relative states—from things being capable of change. Relatives are sweet and bitter and similar concepts; relative states are for instance the right-hand neighbour the father, and similar concepts. . . . In order to explain things more clearly: they call relatives concepts which, being in certain states defined by their own character, are related one to another. But by relative states they mean things which might or might not exist without any internal change and have to be regarded in relation to something outside them. Hexis and knowledge and perception for instance are things that can be in varying conditions each related to the other, and are therefore relative. But when things are compared not by their inherent differences but by their mere relation to something else, they are relative states. The son and the right-hand neighbour stand in need of something outside for their actual existence. For without any change occurring within themselves there will be an end to the state of father if the son dies, and to the state of the right-hand neighbour if the man next to him changes his position. But the sweet and bitter will only change along with their own intensity.

Chapter II

Heracleitos, fragment 67a (Diels)

In the same way as a spider in the centre of the web senses it when a fly destroys one of its threads and quickly hurries to the spot as if it were afflicted by the laceration of the thread, so does the soul of a man, if some part of his body is injured, move quickly thereto as if it were indignant over the injury done to the body with which it has a firm and proper connection.

Chalcidius, *Ad Timaeum*, 220 (Arnim, II, 879)

In the same way as a spider in the centre of the web holds in its feet all the beginnings of the threads, in order to feel by close contact if an insect strikes the web, and where, so does the ruling part of the soul, situated in the middle of the heart, check on the beginnings of the senses in order to perceive their messages from close proximity.

Alexander Aphrodisiensis, *De anima*, 130, 14

Some people explain vision by the stress of air. The air adjoining the pupil is excited by vision and formed into a cone which is stamped on its base by an impression of the object of vision, and thus perception is created similar to the touch of a stick. . . . And with regard to the assumption that vision originates in the *hegemonikon*, if here, too, the theory of tensional motion holds, as they maintain, would there not occur certain interruptions of seeing, since tension would not be produced continuously on the boundaries and neither, consequently, impact? The same question could be asked about what happens when we touch other bodies. For here no interruptions of perception occur, as ought to happen. For such is their theory of tensional motion. If they would argue —as they actually do not—that only the pneuma which they call vision (and not the other senses) performs the said movement called tensional, it would be absurd. On the whole the doctrine of tensional motion is full of paradoxes. This motion represents a uniform primary self-mover which has been shown impossible by those examining the simple movements.[1] Further, if there is one thing which holds together the whole cosmos with all its contents, and for every single and particular body there is something which holds it together, would that not necessarily imply that the same thing carries out simultaneously opposite motions? . . . Moreover, granted that vision originates through a pressure in the air between (the eye and the object), that would obviously lead to the perception of hardness and softness, roughness and

[1] Cf. Arist., *Phys.*, VIII, v.

124

smoothness, moisture and dryness rather than to that of colour. Those (qualities) we can perceive also by a stick, but not colours or shapes or quantity nor anything connected with magnitude, all of which are visible things. But it is colours and shapes and magnitude in particular which are perceived by vision. On the whole this would mean that vision is some kind of touch.

Further: why can objects in the dark not be seen from a place in the light, while from the dark one can see objects which are in the light? For can one believe the doctrine that illuminated air becomes more powerful because of the mixture and can propagate the sensation through pressure, whereas dark air is slack and cannot be stressed by vision, in spite of being denser than illuminated air? . . . It is further a fact that two illuminated houses standing opposite each other can be seen by their inhabitants not less even though the air between them is dark and therefore not stressed.

Sextus Empiricus, *Adv. math.*, VII, 228

Presentation, according to them, is an impression on the soul. But about this they at once began to quarrel; for whereas Cleanthes understood "impression" as involving depression and protrusion, just as does the impression made in wax by signet-rings, Chrysippos regarded such a thing as absurd. . . . He . . . suspected that the term "impression" was used by Zeno in the sense of "modification", so that the definition runs like this— "presentation is a modification of the soul"; for it is no longer absurd that, when many presentations co-exist in us at the same moment, the same body should admit of innumerable modifications; for just as the air, when many people are speaking simultaneously, receives in a single moment numberless and different impacts and at once undergoes many modifications also, so too when the *hegemonikon* is the subject of a variety of images it will experience something analogous to this. (R. G. Bury.)

Sextus Empiricus, *Adv. math.*, VII, 372

If the presentation is an impression on the soul, it is an impression either by way of depression and protrusion, as Cleanthes and his School believe, or by way of mere modification, as Chrysippos thought. And if it subsists by way of depression and protrusion, those absurd results will follow which are alleged by

125

Chrysippos. If the soul when affected by a presentation is impressed like wax, the last motion will always keep overshadowing the previous presentation, just as the impression of the second seal is such as to obliterate that of the first. But if this be so, memory is abolished. . . . (R. G. Bury.)

Sextus Empiricus, *Pyrrh. Hyp.*, II, 70

They say that "presentation" is an impression on the ruling part of the soul. Since, then, the soul, and the ruling part, is pneuma or something more subtle than pneuma, as they affirm, no one will be able to conceive of an impression upon it either by way of depression and prominence, as we see in the case of seals, or by way of the magical "modification" they talk about; for the soul will not be able to conserve the remembrance of all the data that compose an art, since the data which existed before are obliterated by the subsequent modification. (R. G. Bury.)

Empedocles, fragment 109, Diels (quoted in Aristotle, *Metaphysica*, 1000 b 6)

We see Earth by means of Earth, Water by means of Water, divine Aether by means of Aether, destructive Fire by means of Fire, Affection by means of Affection, Hate by means of grievous Hate. (K. Freeman.)

Cicero, *De nat. deor.*, II, 83

But if the plants fixed and rooted in the earth owe their life and vigour to nature's art, surely the earth herself must be sustained by the same power. . . . Her exhalations give nourishment to the air, the aether and all the heavenly bodies. Thus if earth is upheld and invigorated by nature, the same principle must hold good of the rest of the world, for plants are rooted in the earth, animals are sustained by breathing air, and the air itself sees together with us, hears together with us, and utters sounds together with us, since none of these actions can be preformed without its aid. (H. Rackham.)

Sextus Empiricus, *Adv. math.*, VII, 93

And as Poseidonios says in his exposition of Plato's *Timaios*, "Just as light is apprehended by the light-like sense of sight, and sound by the air-like sense of hearing, so also the nature of all things ought to be apprehended by its kindred reason." (R. G. Bury.)

Galen, *De placit. Hippocr. et Plat.* (ed. Müller, p. 641)

That we see through the medium of air is obvious and agreed by all. However, the question is whether this happens as if something comes to us along a path from the object, or if the air serves us as an organ of cognition for an object seen just as the nerves do for an object touched. Most people believe that the modification produced by the things that reach us is transmitted by the nerve to the ruling part of the soul, and thus the object is perceived. But these people do not realize that the sensation of pain could not arise in a limb cut or broken or burned, if the limb itself did not hold the faculty of sensation. But actually the opposite of their opinion is true. The nerve itself namely is part of the brain, like a branch or offshoot of a tree, and this part into which the nerve is rooted fully receives the faculty of sensation and thus becomes able to distinguish the object touching it. Something similar happens to the air surrounding us. When illuminated by the sun it becomes an organ of vision precisely as the pneuma arriving (in the eye) from the brain, but before the illumination occurs which produces a modification through the incidence of the sun's rays the air cannot become such an affected organ.

Galen, *De placit. Hippocr. et Plat.* (ed Müller, p. 625)

Since of all the sensations only the sense of vision receives the perception of the object by transmission through the medium of air, not as from a stick but as some part kindred to and coalescing with it, and only by this and by the incidence of light does it get the speciality of seeing, it stands to reason that it needs the luminous pneuma which flows from above, strikes the surrounding air and assimilates it to itself.

Philo, *Quod deus sit immutabilis*, 35

He endowed some bodies with hexis, some with physis, some with psyche and some with an intelligent soul. In stones and wood which are detached from organic growth he formed hexis as a very strong bond. This is the pneuma that returns upon itself. It begins in the centre of the body and extends outwards to its boundaries, and after touching the outermost surface it turns back till it arrives at the same place from which it started.

Nemesios, *De nat. hom.*, ch. 2

There are those like the Stoics who say that there is a tensional motion in bodies which moves simultaneously inwards and outwards. The outward movement gives rise to quantities and qualities, while the inward movement produces unity and essence.

Alexander Aphrodisiensis, *De mixtione*, 224, 23

What exactly is the simultaneous motion (of the pneuma) in opposite directions by which it makes all bodies containing it coherent and which they call "pneuma moving simultaneously outwards and inwards"? And to what kind of movement does it belong? For in no way can one conceive anything moving by itself simultaneously in two opposite directions.

Cleomedes, *De motu circulari corporum caelestium*, I, 1 (pp. 4, 6, 8)

Outside the cosmos there is the void, extending from all sides into the infinite. That which is occupied by a body is called place, and that which is not occupied is void. Every body must necessarily be in something. That in which it is must be different from the body occupying and filling it, and thus is incorporeal and so to speak insipid. That reality which can receive the body and be occupied by it we call the void. . . .

We can indeed imagine the cosmos moving from the place it occupies now. By this transition we must understand that the place it left becomes void and that which it enters becomes occupied by it; this would be a void which has been filled up. Moreover, if the whole material world is dissolved into fire, as most accomplished physicists believe, the cosmos must necessarily occupy a place which is more than a thousand times greater, like the solid bodies which get vaporized. The place, therefore, which in the state of conflagration will be occupied by the spreading matter is now void and not filled up by any body. . . . When the material world contracts again and is forced into a smaller size, a void will be created again. . . . That which can be filled or left by a body is a void. Therefore, the void must necessarily have some sort of reality.

Such a void, however, does absolutely not exist within the

cosmos. This is evident from the phenomena. For if the whole material world were not completely grown together, the cosmos as it is could not be kept together and administered by nature, nor would there exist a mutual sympathy of its parts; nor could we see and hear if the cosmos were not held together by one tension and if the pneuma were not cohesive throughout the whole being. For the empty spaces in between would impede sense-perception.

Chapter III

Sextus Empiricus, *Adv. math.*, IX, 200-203

And further, since many things become and perish, increase and decrease, move and cease from movement, one must necessarily allow that there exist some causes of these things—some of becoming, others of perishing; some of increase, others of decrease; and also of motion and want of motion. Moreover, even if these effects do not really exist but merely appear, the existence of their causes is introduced once more; for there must exist some cause of their appearing to us as really existing things and not being such.

Again, if there is no Cause all things will have to come from everything and in every place, and also at every time. But this is absurd; for indisputably, if there is no causal law, there is nothing to prevent a horse being formed from a man. And if there is nothing to prevent this, a horse will some time be formed from a man, and likewise, per chance, a plant from a horse. And for the same reason it will not be impossible for snow to congeal in Egypt and drought to occur in Pontos and things proper to summer to happen in winter and things proper to winter to take place in summer. Hence, if what has for its consequence something impossible is itself also impossible, and many impossible consequences follow from the non-existence of Cause, one must declare that the non-existence of Cause also is a thing impossible. (R. G. Bury.)

Sextus Empiricus, *Pyrrh. hyp.*, III, 17

That Cause exists is plausible; for how could there come about increase, decrease, generation, corruption, motion in general, each of the physical and mental effects, the ordering of the whole universe, and everything else, except by reason of some cause?

For even if none of these things would exist in accordance with nature, we should affirm that it is due to some cause that they appear to us other than they really are. (R. G. Bury.)

Alexander Aphrodisiensis, *De fato*, ch. 22 (p. 191, 30 f.)

They say that this cosmos is one and comprises in it all that exists and is administered by a vital, reasonable and intellectual principle of growth. It holds the eternal administration of all that exists which advances by a certain sequence and order. The prior events are causes of those following them, and in this manner all things are bound together with one another, and thus nothing can happen in the cosmos which is not a cause to something else following it and linked with it. Nor can any one of those events happening later be separated from what happened earlier and not be tied up with one of these earlier happenings. But from everything that happens something else follows depending on it by necessity as cause. . . . For there neither exists nor happens anything uncaused in the cosmos, because there is nothing in it which is separated and divorced from all that happend before. The cosmos would break up and be scattered and could no longer remain a unity administered by one order and plan, if some uncaused movement were to be introduced into it. This would be the case if there were not some cause preceding it for all that exists and happens, from which it results by necessity. According to them a causeless event is equal in essence to and equally impossible as a creation out of nothing. Such is the administration of the All going on from infinity to infinity manifestly and unceasingly. . . .

In view of the multiplicity of causes, the Stoics equally postulate about all of them that, wherever the same circumstances prevail with regard to the cause and the things affected by the cause, it is impossible that sometimes the result should be this and sometimes that; otherwise there would exist some uncaused motion.

Cicero, *De fato*, 40-43

Let us if you please examine the nature of Chrysippos' doctrine in connection with the topic of assent (to sense-presentation) which I treated in my first discourse. Those old philosophers who held that everything takes place by fate used to say that assent is given perforce as the result of necessity. On the other hand those who disagreed with them released assent from bondage to fate, and maintained that if assent were made subject to fate it would be impossible to dissociate it from necessity[1]. . . . But Chrysippos, since he refused on the one hand to accept necessity and held on the other hand that nothing happens without fore-ordained causes, distinguishes different kinds of causation, to enable himself at the same time to escape necessity and to retain fate. "Some causes", he says, "are perfect and principal, others auxiliary and proximate. Hence when we say that everything takes place by fate owing to antecedent causes, what we wish to be understood is not perfect and principal causes, but auxiliary and proximate causes." Accordingly he counters the argument that I set out a little time ago by saying that, if everything takes place by fate, it does indeed follow that everything takes place from antecedent causes which are not principal and perfect but auxiliary and proximate causes. And if these causes are not in our power, it does not follow that desire also is not in our power. On the other hand if we were to say that all things happen from perfect and principal causes, it would then follow that, as those causes are not in our power, desire would not be in our power either. . . . But Chrysippos goes back to his cylinder and spinning-top, which cannot begin to move unless they are pushed or struck, but which when this has happened, he thinks, continue to move of their own nature, the cylinder rolling and the top spinning round. "In the same way therefore," he says, "as a person who has pushed a cylinder forward has given it a beginning of motion, but has not given it the capacity to roll, so a sense-presentation when it impinges will it is true impress and as it were seal its appearance on the mind, but the act of assent will be in our power, and as we said in the case of the cylinder, though given a push from without, as to the rest will move by its own force and nature. If some event were produced without ante-cedent cause, it would not be true that all things take place by fate; but if it is probable that with all things whatever that take place there is an antecedent cause, what reason will it be possible to adduce why we should not have to admit that all things take place by fate? Only provided that the nature of the distinction and difference between causes is understood." (H. Rackham.)

[1] Necessity is used here in the sense of negation of free will.

Nemesios, *De natura hominis*, ch. 35

There are people who say that both free will and fate can be preserved. For to everything that happens something is ordained by fate, for example being cold to water, bearing fruit to a plant, falling to stones and rising to fire, and to living beings to assent and to act by impulse; and whenever none of the external things given by fate resists this impulse, walking will be completely by our free will and we surely will walk. Those who maintain this (among the Stoics Chrysippos and Philopator and many other eminent men) do not prove anything else but that everything happens by fate.

For if the impulses are given us by fate and they are sometimes hindered by fate and sometimes not, it is obvious that everything will happen by fate including that which seems to be in our power. We will again use the same reasoning against them and show the absurdity of their doctrine. For if in the case where the same circumstances prevail everything must happen exactly in the same way, as they say, and not sometimes one way and sometimes another because it was so allotted since eternity, then the impulse of the living being, too, must necessarily and absolutely arise in the same way if the same causes prevail. But if the impulse follows by necessity, where remains our free will? . . . There would only be a free will, if under the same circumstances it were in our power sometimes to follow our impulse and sometimes not. . . . The same will be found true for animals and inanimate things. For if they (the Stoics) assume that the impulse is in our power because we possess it by nature, why should one not say about fire that it burns by its free will because it is its nature to burn—which Philopator seems to have maintained somewhere in his book *On Fate*? However, that which happens *through* us by fate does not happen *by* us. By the same reasoning it should be the free will of the lyre or the flute and the other instruments and of all the animals and inanimate things, if somebody operates through them—which is absurd.

Alexander Aphrodisiensis, *De fato*, ch. 13 (p. 181, 13 ff.)

The Stoics say that "*in* our power" is that which happens "*through* our power", thus denying that man has the power to choose and to do one of opposites. For, they argue, as the nature of things that are and happen is various and different, . . . all that happens to any thing will happen in accordance with its specific nature. What happens to a stone will happen in accordance with the nature of the stone, and so with fire and with a living being. However, none of these happenings in accordance with their specific nature could happen otherwise, but they all happen so by a necessity and not by force, out of the impossibility of what is natural to move in one case thus and in another differently. For if one lets a stone fall from a certain height it must move down if there is no impediment, because it contains gravity in itself which is the cause of its natural movement. And if there is also present the external cause which assists the stone in its natural movement, the stone must by necessity move according to its nature. . . . The same law applies to other things, and what holds for inanimate things holds also for living beings. For there exists a certain natural movement in living beings—the movement by impulse. For every living being which moves as a living being carries out a movement originating in it according to fate as a result of an impulse. Under these circumstances movements and actions go on in the cosmos according to fate, some for instance through the medium of earth, some through air, some through fire and some through something else, and in the same way some are effected by living beings . . . and these, happening according to fate, are said by the Stoics to happen by free will of those living beings.

Diogenianos (Eusebios, *Praep. Evang.*, IV, 3, 1)

Chrysippos gives another proof in the above mentioned book which runs as follows: according to him the predictions of the diviners could not be true if fate were not all-embracing. This statement is very naïve; as if it were clear that all the predictions of the so called diviners came true, or, moreover, as if someone, without telling a lie, would concede that everything happens according to fate, for the evidence proves the contrary, namely that not all predictions and not even most of them come true.

Chrysippos has given us a proof based on the mutual dependence of things. For he wants to show by the truth of divination that everything happens in accordance with fate; but he cannot prove the truth of divination without first assuming that everything happens in accordance with fate. Could there be a more fallacious way of proving it? For if something turns out evidently as predicted by the diviners it would not be a sign that divination is a science but that events coincide by chance with predictions. This does not prove any scientific truth. We could not call somebody a skilful archer who once happens to hit the mark but mostly misses it, nor a physican who kills most of his patients and only occasionally saves one. Nor would we call it skill at all if we did not succeed in all our undertakings or at least in most of them. . . . That the so called diviners sometimes prove correct in their predictions is not the result of science but of a chance-like cause. Being careful about the clear definition of terms we call a case where somebody succeeds a chance-like event, not when the set goal has never been attained, but when it has neither been attained in all cases nor in most cases, nor as a consequence of scientific knowledge.

Cicero, *De divinatione*, I, 24-25

But (say the opponents of divination) sometimes predictions are made which do not come true. And pray what art does not have the same fault? By art I mean the kind that is dependent on conjecture and deduction. Surely the practice of medicine is an art, yet how many mistakes it makes! . . . And is military science of no effect because a general of the highest renown recently lost his army and took to flight? . . . So it is with the responses of the diviners from the entrails of animals and, indeed, with every sort of divination whose deductions are merely probable; for divination of that kind depends on inference and beyond inference it cannot go. It is sometimes misleading perhaps, but none the less in most cases it guides us to the truth. For divination since times immemorial has grown into an art through the repeated observation and recording of almost countless instances in which the same results have been preceded by the same signs. (W. A. Falconer.)

loc. cit. I, 127-128

Moreover, since, as will be shown elsewhere, all things happen by fate, if there were a man whose soul could discern the links that join each cause with every other cause, then surely he would never be mistaken in any prediction he might make. For he who knows the causes of future events necessarily knows what every future event will be. But since such knowledge is possible only to a god, it is left to man to presage the future by means of certain signs which indicate what will follow them. Things which are to be do not suddenly spring into existence, but the evolution of time is like the unwinding of a cable: it creates nothing new and only unfolds each event in its order. . . . Therefore it is not strange that diviners have a presentiment of things that exist nowhere: for all things *are*, though, from the standpoint of time they are not present. As in seeds there inheres the germ of those things which the seeds produce, so in causes are stored the future events whose coming is foreseen by reason or conjecture, or is discerned by the soul when inspired by frenzy, or when it is set free by sleep. Persons familiar with the rising, setting and revolutions of the sun, moon and other celestial bodies can tell long in advance where any one of these bodies will be at a given time. And the same thing may be said of men who, for a long period of time, have studied and noted the course of facts and the connexion of events, for they always know what the future will be; or, if that is putting it too strongly, they know in a majority of cases; or, if that will not be conceded either, then, surely, they sometimes know what the future will be. (W. A. Falconer.)

Boethius, *In De interpretatione*, II, p. 194

The Stoics who believe that everything happens out of necessity and by providence, judge the casual event not according to the nature of chance itself but according to our ignorance; for they take as casual that which, though happening by necessity, is not known to men.

Alexander Aphrodisiensis, *De anima*, 179, 6

The assertion that chance is a cause obscure to the human mind is not a statement about the nature of chance but means that chance is a specific relation of men towards cause, and thus the same event appears to one as chance and to another not, depending on whether one knows the cause or does not know it. . . . If they mean to define chance as obscure to those who are ignorant of the cause, then by the same definition all processes of science and arts would be chance to the ignorant and unskilled. For a man who is no carpenter does not know the rules of carpentry, neither does somebody who is not a musician know the rules of music, nor any other unskilled the rules of the art. For skill means knowing the causes of the processes of art.

Alexander Aphrodisiensis, *De fato*, 176, 14

Their theory is that the possible and contingent does not exclude that everything shall happen according to fate, and they define the possible event as something that is not prevented by anything from happening even if it does not happen. "There is nothing to prevent the occurrence even of the opposite of what happens through fate, for even though it does not occur it is still possible", and the fact that the preventing causes are not known to us is the reason for the assumption that there was no hindrance for the things to happen. For these things which are the causes for the opposite things to happen according to fate are also the causes for the non-happening of the things themselves, if, as they say, it really is impossible that under the same circumstances the opposite should happen. But because we have no knowledge of things which happen, therefore, so they say, things which do not happen seem possible to us—is not this view ridiculous?

Simplicios, *In Categ.*, 195, 32

How shall we decide whether something is perceptible or knowable? Shall we decide by the fitness[1] alone, as Philo said, even if there is no knowledge of it, nor ever will be? As for instance that a piece of wood in the Atlantic Ocean is combustible in itself and according to its own nature? Or shall it rather be decided by the unhindered fitness according to which a thing can be subject to knowledge and perception by itself as long as no manifest obstacle stands in the way? Or by neither of these two, and shall that be called knowable if knowledge of it either exsits or will exist, and the possible will be decided by the outcome?

Plutarch, *De Stoic. repugn.*, 1055 d

Is it not clear that Chrysippos' doctrine of the possible contradicts his doctrine of fate? For if the possible is not what either is or will be true, as Diodoros states, but if everything is possible that admits of happening even if it will not happen, then many

---

[1] In the sense of capability of coming into being.

of those things will be possible which according to insuperable and unviolable and victorious Fate will not happen. Thus either fate's power will dwindle, or, if fate is that what Chrysippos believes it to be, that which admits of happening will often become impossible.

Boethius, *In De interpretatione*, II, p. 234

Philo says that something is possible if in its internal nature it is susceptible of truth, as for instance if I say that today I shall read again Theocritos' *Bucolica*. This, if nothing external prevents it, can be truly stated, as far as it concerns its inner nature . . . Diodoros defined as possible that which either is or will be.

Cicero, *De fato* 13

This is a view that you, Chrysippos, will not allow at all, and this is the very point about which you are specially at issue with Diodoros. He says that only what either is true or will be true is a possibility, and whatever will be, he says, must necessarily happen and whatever will not be, according to him cannot possibly happen. You say that things which will not be are also possible—for instance it is possible for this jewel to be broken even if it never will be. (H. Rackham.)

Chapter IV

Clemens Alexandrinus, *Stromateis*, VIII, 9

There are things where one is not the cause of the other but where they are causes mutually depending on each other. For instance a given condition of the spleen is not the cause of fever but of the beginning of fever, and a given fever is not the cause of the spleen but of a change in its condition. In the same way virtues are mutual causes such that they cannot be separated because of their interdependence. And the stones of a vault are each other's cause for staying put, but one is not the cause of the other. Pupil and teacher are mutual causes for progress. Sometimes one speaks of mutual causes of the same thing, for instance the importer and the retail trader who are mutual causes for making profit; and sometimes of different things, for instance the knife and the meat: the one is the cause for the meat to be cut, the other the cause for the knife to cut.

Simplicios, *In Categoriam*, 237, 29

They say that hexis can be tightened and loosened but dispositions are neither capable of increase nor of diminution. Therefore they call the straightness of a rod a disposition although it can easily be changed by bending. For straightness cannot be loosened nor stretched, nor does it admit of variation in degree because it is a disposition. Similarly virtues are dispositions, not because of their specific stable properties but because they cannot be intensified and do not admit of increase. But skill, even if it cannot vary easily, is not a disposition. They seem to understand hexis as the range of variation of a state and disposition as the extreme case, whether it, like the straightness of a rod, can easily change or not.

Plutarch, *Plat. quaest.*, V

Moreover, the smaller a thing is, the nearer it is to the fundamentals; and the straight line is the shortest of all lines. A circle, however, is concave with respect to its interior and convex with respect to its exterior. . . . Moreover, a straight line, be it short or long, maintains its straightness, whereas the curvature of circles increases with their smallness and vice versa.

Diogenes Laertius, VII, 72

A proposition which indicates more is one that is formed by the word signifying "more" and the word "than" in between the clauses, as for example "It is more day than night." A proposition which indicates less is the opposite of the foregoing, as e.g. "It is less night than day." (R. D. Hicks.)

Simplicios, *Phys.*, 136, 10

In connection with the same theory of dichotomy, Alexander says, . . . Xenocrates the Chalcedonian has shown that the divisible Whole consists of many parts (because the part differs from the whole) and that the same thing cannot be one and at the same time many because contradicting things cannot be true together, but that it is not true that every quantity is divisible and has parts. For there exist certain atomic lengths for which it never is true that they consist of many parts. Thus he believed to have found the nature of the One and to have escaped the contradiction, because on the one hand the divisible is not one but many, and on the other hand the atomic lengths are not many but are only one.

Philoponos, *Phys.*, 83, 19

With Zeno's theory Xenocrates is said to have taken issue by assuming that the division of quantities does not go on infinitely, and that the divided line ends in atomic lengths. He did not realize that he had got caught in a contradiction by imagining he had escaped from another one. For it is not impossible for one thing to be one and many, nor is there any contradiction if one is taken potentially and the other actually. But to assume that a line and an indivisible are one and the same means obviously to assume a line which is not a line and a quantity which is not a quantity, if really a quantity is an infinitely divisible thing.

Philoponos, *Phys.*, 84, 15

Those who assumed falsely that quantities are not divisible ad infinitum wrongly yielded to Zeno's paradox by which he proved the oneness and immobility of the existing by the infinite division of quantities. They declare that if quantities were infinitely divisible there would be no motion nor anything essentially one and therefore also not many, because plurality is composed of many units. Therefore Xenocrates denied the infinite division of quantities. He asserted the truth of this in many ways but refuted the paradox in one way by saying that the division of quantities is only potentially infinite, but not actually (because it is impossible for the infinite to become actual). In this way motion, though potentially passing through an infinity of points, will actually pass through a finite number. For motion does not take place in relation to points but to definite quantities.

Sextus Empiricus, *Adv. math.*, IX, 419

The Geometers are of the opinion that the straight line by revolving describes circles with all its parts. But the fact that the line is length without breadth conflicts at once with this theorem of theirs. For every part of the line, as they assert, contains a point, and the point by revolving describes a circle. Thus when the straight line, by revolving and describing a circle with all its parts, has swept out the area which extends from the centre to the outermost circumference, then the parallel circles are either continuous or separate from one another. But whichever of these alternatives the Geometers may adopt, they will involve themselves in an almost insuperable difficulty. For if these circles are separate from one another there will be a certain part of the area which has not been swept out by a circle, and of the line which has not described a circle, namely that which is situated at this interval of the area. But this is absurd; for the line certainly contains at this part a point, and the point by revolving at this part describes a circle, for that the line at any part of it should not contain a point or that the point should not by revolving describe a circle, is contrary to the Geometers' doctrine. And if the circles are continuous, either they are continuous in such a way as to be situated in the same place or so that they are conceived as lying side by side in such a way that no point can be inserted between them; for if one is inserted, it is bound to describe a circle. And if they occupy the same place they will all become one, and because of this the greatest circle will not differ from the least . . . which is contrary to sense. (R. G. Bury.)

Plutarch, *De communibus notitiis:*
1080 c

There exists a quantity which is greater than but not exceeding another one.

1079 e

Now see how Chrysippos answered Democritos who raised the following pertinent question in natural philosophy: If a cone were cut by a plane parallel to the base, what must we think of the surface of the sections? Are they equal or unequal? For if they are unequal, they will make the cone irregular, as having many indentations, like steps, and unevennesses; but if they are equal, the sections will be equal, and the cone will appear to have the property of the cylinder and to be made up of equal, not unequal circles, which is quite absurd. Thereupon he enlightened Democritos' ignorance as follows: The surfaces will neither be equal nor unequal; the bodies, however, will be unequal, since their surfaces are neither equal nor unequal.

1079 d

The sides of the triangle of which a pyramid is made up (by parallel sections), being inclined towards the apex, although they are unequal, do not exceed one another, even if one is greater than the other.

1078 e

There is no extreme body in nature, neither first nor last, into which the size of a body terminates. But there always appears something beyond the assumed, and the body in question is thrown into the infinite and boundless.

1080 e

What the Stoics bring forward most strongly against the champions of the indivisibles is this: "There is no contact between bodies as a whole nor part by part. The reason for the first is that they make no contact but mix one with another, and the reason for the latter is that this would be impossible because indivisibles have no parts."

1079 a

Is it not evident that man consists of more parts than his finger, and again that the cosmos consists of more parts than man? Everyone takes it for certain and believes it, except the Stoics. The Stoics say and believe the contrary, i.e. that "man does not exist of more parts than his finger, nor the cosmos of more parts than man. For the division of bodies goes on infinitely, and among the infinites there is no greater and smaller nor generally any quantity which exceeds the other. Nor cease the parts of the remainder to split up and to supply quantity out of themselves."

1079 b

If we are asked whether we have any parts and how many, and of what and of how many parts these consist, we will have to make a distinction. On the one hand we can posit large parts and say that we are composed of head and trunk and limbs— this was all that was asked and inquired about. On the other hand, if the question is carried further to the least parts, nothing of this kind can be assumed, but one must say that we are neither composed of such and such parts nor of so many, nor of finite or infinite ones.

1081 c

The Stoics do not admit the existence of a shortest element of time, nor do they concede that the Now is indivisible, but that which someone might assume and think of as present is according to them partly future and partly past. Thus nothing remains of the Now, nor is there left any part of the present, but what is said to exist now is partly spread over the future and partly over the past.

1081 f

Chrysippos in the third, fourth and fifth book of his work *On Parts* declares that part of the present time is future and part past.

Stobaios, *Eclogae:*

I, 104

Zeno said that time is the interval of movement which holds the measure and standard of swiftness and slowness.

I, 106

Chrysippos defined time as interval of movement which sometimes is also called measure of swiftness and slowness, or the interval proper to the movement of the cosmos. And it is in time that everything moves and exists. It seems that time is to be taken in two senses, just like the earth and the sea and the void, namely in the sense of the whole and its parts. In the same way as the void is all infinite everywhere, so time is all infinite in both directions; indeed, past and future are both infinite. And he states most clearly that no time is entirely present; for the division of continua goes on infinitely, and by this distinction time, too, is infinitely divisible. Thus no time is strictly present, but is defined only loosely.

I, 105

Appollodoros in his book on physical science defines time thus: Time is the interval of movement of the cosmos. It is thus infinite, as number as a whole is said to be infinite. Part of it is past, part present, part future; and the whole time is passing just as we say that the year passes on a larger circuit. And time is said to be really existing while none of its parts exists precisely.

Plutarch, *De Stoic. repugn.:*

1052 c

In the first book of his work *On Providence* Chrysippos says: "Zeus grows till he consumes everything. As death is the separation of soul from body and as the soul of the cosmos does not leave it but is growing continuously until it has consumed all matter, one cannot say that the cosmos is mortal."

1053 b

In the first book of his work *On Providence* Chrysippos says: "When the Cosmos is completely in the fiery state, so, at the same time, are its soul and its *hegemonikon*. But if what is left over of the soul is changing into the humid state, the cosmos is in a certain way transformed into body and soul, and thus, composed of both, it exhibits another order." From this it is obvious that according to Chrysippos the lifeless parts of the cosmos, too, become animated in the state of conflagration. But in the cooling-down period the (cosmic) soul slackens again and becomes more humid, thus passing into a body-like state.

Cleomedes, *De motu circulari corporum caelestium*, I, 1 (p. 10 f)

But, they (the Peripatetics) say, if outside the cosmos were a void, the cosmos would move through it as nothing would be able to hold it together and to support it. To this we reply that this would not happen because the cosmos tends towards its own centre, and this centre is situated below, wherever the cosmos tends. For if the centre and the "below" of the cosmos were not the same, the cosmos would move downwards through the void, as will be shown in the discourse on the centripetal motion.

They also argue that if there existed a void outside the cosmos, matter would be poured into it and dispersed and scattered into infinity. Our reply is that this could never happen, because the cosmos has a hexis which holds it together and protects it, and the surrounding void cannot affect it. It maintains itself by the rule of an immense force, contracting and again expanding into the void according to its natural transmutation, alternately dissolving into fire and starting creation again.

That a void outside the cosmos is necessary is evident by what has been shown above. However, that it extends from all sides to infinity is of the utmost necessity, as can be seen from that which follows. Everything finite is bordered by something of a different kind. . . . It would, therefore, be necessary, if the void surrounding the cosmos were finite, for it to terminate into something of a different kind. But there does not exist anything differing in kind from the void into which it terminates and therefore it is infinite.

Simplicios, *Phys.*, 671, 4

Alexander says that one can apply the same argument against the Stoic theory of the infinite void surrounding the cosmos. For why does the cosmos remain at rest where it is and not move, if the void is infinite? And further, if it moves, why rather there than elsewhere? For the void is indistinguishable and equally yielding everywhere. But if they say that the cosmos remains at rest because of the hexis which holds it together, then one might well concede that the hexis could contribute something to prevent the dispersion and scattering of the parts of the cosmos in different directions. But to keep the whole including the cohesive hexis at rest and prevent its motion—this the hexis can not do. It would have to be proved by them, Alexander says, that the void can thus not be a "material" cause of movement if the cosmos put into it has no cause to move in either direction.

Plutarch, *De Stoic. repugn.*:
1054 f

If the whole universe is thus directed and moving towards itself and if its parts possess this movement by the nature of the body, it is plausible that all bodies possess a primary natural movement towards the centre of the cosmos; the cosmos thus moves towards itself and its parts *qua* parts.

1054 f

The cosmos is a perfect body but its parts are not perfect, as they exist only relative to the whole and have no absolute existence.

1052 d

Of the cosmos alone it can be said that it is self-supporting, because it alone contains within itself all that it needs. It is fed and grows out of itself, whereas its parts are in mutual exchange with each other.

1056 e

There are many obstacles and impediments for partial entities and movements, but none for the whole.

Marcus Aurelius, *Medit.*, II, 9

This thou must always bear in mind, what is the nature of the whole, and what is my nature, and how this is related to that, and what kind of a part it is of what kind of a whole; and that there is no one who hinders thee from always doing and saying the things which are according to the nature of which thou art a part. (G. Long.)

SELECTED BIBLIOGRAPHY

SOURCES

ALEXANDER APHRODISIENSIS, *Scripta minora*, ed. I. Bruns, vols. I and II; Berlin, Reimer, 1887 and 1892. (Contains i.a. *De anima*, *De fato*, *De mixtione*.)

ARNIM, J. v., *Stoicorum veterum fragmenta*, vols. I-IV; Leipzig, Teubner, 1921.

BOETHIUS, *Commentarii in librum Aristotelis De interpretatione*, ed. C. Meiser; Leipzig, Teubner, 1880.

CICERO, *De natura deorum*; *Academica*; *De fato*, with an English translation by H. Rackham, Loeb Classical Library; London, Heinemann, 1948, 1951.

—— *De divinatione*, with an English translation by W. A. Falconer, Loeb Classical Library; London, Heinemann, 1953.

CLEMENS, ALEXANDRINUS, *Stromateis*, Patrologia Graeca; Paris, Garnier, 1890, vol 9.

CLEOMEDES, *De motu circulari corporum caelestium*, ed. H. Ziegler; Leipzig, Teubner, 1891.

DIELS, H. *Die Fragmente der Vorsokratiker*, 6th ed.; Berlin, Weidmannsche Verlagsbuchhandlung, 1951.

DIOGENES LAERTIUS, *Vitae philosophorum*, with an English translation by A. D. Hicks, Loeb Classical Library; London, Heinemann, 1950, 2 vols.

EUSEBIOS, *Praeparatio evangelica*, ed. K. Mras; Berlin, Akademie-Verlag, 1954.

GELLIUS, AULUS, *Noctes Atticae*, with an English translation by J. C. Rolfe, Loeb Classical Library; London, Heinemann, 1948, 3 vols.

NEMESIOS, *De natura hominis*, Patrologia Graeca; Paris, Garnier, 1886, vol. 40.

PHILOPONOS, Ioannes, *In Artistotelis Physicorum libros commentaria*, ed. H. Vitelli; Berlin, Reimer, 1887.

PLUTARCH, *Moralia*, ed. F. Dübner; Paris, Didot, 1841: ed M. Pohlenz; Leipzig, Teubner, 1952.

SEXTUS EMPIRICUS, *Sextus Empiricus*, with an English translation by R. G. Bury, Loeb Classical Library; London, Heinemann, 1939-1957, 4 vols.

SIMPLICIOS, *In Aristotelis Categorias commentarium*, ed. C. Kalbfleisch; Berlin, Reimer, 1907.

—— *In Aristotelis Physicorum libros commentaria*, ed. H. Diels; Berlin, Reimer, 1882.

OTHER WORKS

BREHIER, E., *Chrysippe et l'Ancien Stoicisme*; Paris, Presses Universitaires, 1951.

GOLDSCHMIDT, V., *Le Système Stoicien et l'idée de Temps*; Paris, J. Vrin, 1953.

MATES, B., *Stoic Logic*; Berkeley, University of California Press, 1953.

POHLENZ, M., *Die Stoa*; Goettingen, Vandenhoek and Ruprecht, 1948.

REINHARDT, K., *Kosmos und Sympathie*; Muenchen, Beck'sche Verlagsbuchhandlung, 1926.

ZELLER, E., *Die Philosophie der Griechen*, ed. Wellmann, 5th ed.; Leipzig, O. R. Reisland, 1923.

INDEX TO PASSAGES QUOTED

INDEX TO PASSAGES QUOTED

GENERAL INDEX

Aether, 34, 37 ff, 119
Aggregate, continuous, 98, 141
Air:
  activity of, 3
  identified with intelligence, 10
  mixture with light, 28
  neutral position of, 3
  non-heavy, 6, 111
  stressed, 23
Alcmaeon, 2, 21
Alexander Aphrodisienses:
  criticism of Stoic theory of mixture, 15
  criticism of Stoic notion of causality, 54
Anaxagoras, 16, 97
Anaximander, 108
Anaximenes, 2, 7
Antiphon, 90
Apollodoros, 106, 142
Archimedes, 87, 91
Archytas, 27, 100
Aristotle:
  his categories, 17
  his classification of causes, 50
  on mixture, 12
  on the possible, 71 f
  on time, 100
Atomic lengths, 91 f, 139

Body, dynamic conception of, 95 f, 141
Boethius, 71, 74 f
Boyle, 35
Brouwer, 98
Bryson, 90

Categories:
  Aristotelian, 17
  Stoic, 17 f
Causal law, 50, 129 f
Causes :
  antecedent, 60, 131
  corporeal nature of, 53, 82, 95
  hidden, 64
  intensifying, 60
  joint, 60
  multiplicity of, 54, 130

mutual, 82, 137
  operative, 60, 83
  principal or perfect, 63
Chance, 56, 76, 135
Chrysippos, 24 f, 44, 69, 78 f, 98, 111, 115
  against causeless motion, 56, 64, 131
  and Democritos' paradox, 93 f, 140
  on free will, 61 ff
  on time, 101 f, 142
Cicero:
  on determinism, 58 f
  on divination, 66 ff
  on the nature of heat, 4, 117
Cleomedes, 41, 110, 143
Cohesion, 1 ff, 31
Cold, active properties, 43
Communication between parts and whole, 8, 81, 114
Comparative proposition, 88, 138
Conditions, initial, 59, 62, 65
Conflagration, 43, 106 f, 128
Conservation, laws of, 55 f
Convergence, 89, 91, 93
Correlatives, 82
Cosmos:
  a closed system, 114
  cycles of, 107, 143
  finiteness of, 1, 43, 108
  hexis of, 113, 143
  tendency towards the centre, 111 f
Curvature of circles, 88

Democritos, 46, 108
  on the paradox of the cone, 92 f, 140
Descartes, 37, 84
Differentiation, principle of, 11, 45
Dilution, 14
Diodoros, 73 f, 78, 136
Diogenes of Apollonia, 10, 26, 116
Diogenianos :
  on free will, 61
  on divination, 68 ff
Disjunction, 76 f
Disposition, 84, 138